D1484332

The
GIZA
POWER PLANT

"None of the previous theories regarding the purpose of the Great Pyramid really explain the known facts. From tombs for the pharaohs, to markers for astrology, every explanation, thus far, has failed to account satisfactorily for at least some aspect of the physical evidence. Christopher Dunn, however, has looked squarely at all the available research and provided a truly convincing and scientific, albeit revolutionary, explanation in *The Giza Power Plant.*"

—Douglas Kenyon, Editor of *Atlantis Rising*

The
GIZA
POWER PLANT

TECHNOLOGIES of ANCIENT EGYPT

Christopher Dunn

BEAR & COMPANY
PUBLISHING
SANTA FE, NEW MEXICO

Library of Congress Cataloging-in-Publication Data

Dunn, Christopher P., 1946–
    The Giza power plant : technologies of ancient Egypt / by
Christopher P. Dunn.
       p.   cm.
    Includes bibliographical references.
    ISBN 1-879181-50-9
    1. Great Pyramid (Egypt).   2. Power plants—Egypt—History.
3. Technology—Egypt—History.   I. Title.
DT63.D86   1998
932—DC21   98-3949
                                                CIP

Bear & Company, Inc.
Santa Fe, NM 87504-2860

Cover art: © 1998 by Bleu Turrell

Cover design: © 1998 by Lightbourne

Interior page design and typography: Beth Hansen-Winter

Hieroglyphic drawings: © 1998 Peter Winter

Editing: Joan Parisi Wilcox

Printed in the United States by R.R. Donnelley

9   8   7   6   5   4   3   2

DEDICATION

To Jeanne, Peter, Alexander, Geno, my parents,
and my European family and friends.

CONTENTS

LIST OF ILLUSTRATIONS

ACKNOWLEDGMENTS

A single name is on the cover of this book because there is not room for each person who should get credit for the work contained within. Ever since I was able to read, I have held authors in high esteem. In fact, I am awestruck by them. Rarely, though, did I read their acknowledgments page, and when I did I did not fully appreciate the importance of the author's credits. Now I do. I now realize that in order to make a book possible, an author needs mountains of support from family, friends, colleagues, and ultimately the editors who guide the rough manuscript to book form. Now that this book is in your hand, and I have joined the ranks of authors, I am humbled by the process that has brought me here, and would like to place credit where it belongs.

Without the support of my family, this book would not be possible. The untold hours on the computer, while chores and family time were put on hold, could only have been spent with the full support and kind patience of my wife, Jeanne. Even before the manuscript found a publisher, Jeanne spent many hours on the old manuscript, which had been started in 1977 and had not been worked on since 1983, creating an electronic copy and fixing my grammar and punctuation. The enthusiasm of my children, who write proudly in their school essays that "Father is an engineer and author," has surprised and delighted me.

Though we have never met, and have only talked on the telephone for a few minutes, I have spent many hours with Peter Tompkins, via his book *Secrets of the Great Pyramid*. Without his efforts in making early research on the Great Pyramid available in a descriptive and readable way, *The Giza Power Plant* would not have been given life.

I started *The Giza Power Plant* in 1977. I gave it the title on the day of its genesis and it will be of age in September 1998. Over the past twenty-one years, I have received guidance, information, advice, and encouragement from many people from all walks of life. Without Paul and Ardith Keller of Camby, Indiana, the work would not have even started. They helped me change my perspective on my life, my spirit, and my soul—they made it possible for me to imagine the unimaginable. I would also like to thank Lexey Ann for introducing me to them.

I would like to thank my dear friends Arlan and Joyce Andrews, who goaded me into submitting parts of my book for publication in magazines—on more than one occasion—and for presenting me with a talented foil against which I honed my ideas. Thanks also to Dr. Stanley Schmidt of *Analog* magazine for seeing the value of "Advanced Machining in Ancient Egypt" and seeing fit to publish it on two occasions, and to Jeff D. Kooistra, for reading *The Giza Power Plant* synopsis and adding his own perspective on Tesla technology.

I greatly appreciate my dear friends, Jeff and Judie Summers, who have given me encouragement and support in many different ways; and Tom Adams, Steven Defenbaugh, Joe Drejewski, Arlen Gondry, Joe Klinger, Dr. Katherine Klinger, Judd Peck, Donald Raha, Max Rettig, Clyde Treadwell, and my friends at the Danville Engineers Club.

In the past three years I have been fortunate to meet other researchers who question orthodox beliefs and who are working tirelessly to bring enlightenment and truth to the world. Each brings a different perspective on ancient cultures, and, while we do not necessarily all agree on every aspect, I have been amazed at the high level of cooperation and support shown by Robert Bauval, Mike Brass of the University of Capetown, South Africa, David Hatcher Childress, James Hagan, Graham Hancock, and Roger Hopkins; Laura Lee, Eric Leither, Robert McKenty, Stephen Mehler, Tom Miller, and Richard Noone; Chris Ogilvie, Roel Oostra, Dr. Robert Schoch, Robert Vawter, John Anthony West, and Colin Wilson. I also appreciate those who have sent me e-mails after reading my article on the Internet. They are too numerous to mention, but should know that I appreciate their comments and support. I also appreciate those who have challenged my ideas, for I find that these challenges are great opportunities for growth and learning. A debate on "Advanced Machining in Ancient Egypt" raged for almost six months on the sci.archaeology newsgroup, and was brought to my attention by Rodney Small, who had read the original article in 1984. I would like to thank the principal debaters on this issue, Miguel Magguire, August Matthusson, and Martin Stower, for their most eloquent and scholarly contribution. It was August who kept pounding on the question, "Where are the power plants?"

J. Douglas Kenyon, editor of *Atlantis Rising*, is largely responsible for bringing contact between me and Barbara Clow of Bear & Company. Barbara tore apart a manuscript that had been lying dormant for so many years that it had calcified. After I submitted it three times, feeling as though I had been brought to the front of the class each time, she finally agreed that it should go to copyediting. Little did I know at this point that the real work was to begin; I cannot thank Joan Parisi Wilcox enough for holding my feet to the fire and the effort she put into the manuscript in making it what it is now. I appreciate her scholarship, her original skepticism, and her subsequent belief in the radical theory *The Giza Power Plant* proposes. Many thanks also to editorial director, John Nelson; publicity director, Jody Winters; and the rest of the Bear clan who make this book possible.

And to the people of the United States of America, who have created an environment that nurtures freedom, creativity, and opportunity. As an immigrant, I fully recognize the benefits this environment has given me.

INTRODUCTION

or years there has been a strong belief that a highly advanced civilization populated this planet thousands of years ago. This belief seems to be increasing and affecting not just the fringes of academic thought, but as the new millennium approaches, more conventional scholars and their students. Anomalous artifacts have been found in Egypt and other places that imply the use of what we would consider advanced technology, by either their function or design and manufacture. Did our distant ancestors possess scientific knowledge and technical skills that we have struggled to acquire for centuries? Many people would emphatically answer "Yes"! Based on logical arguments that reference artifacts from ancient times, scholars and laypeople alike are slowly coming to the realization and giving credence to the idea that cataclysmic forces brought a technologically astute civilization to an end.

Understandably, this movement, which threatens to shake the foundations of Western orthodoxy, has its antagonists. And in rebutting speculations about the existence of technologically advanced civilizations in prehistory, orthodox scholars pose pertinent questions: Where is the infrastructure to support such a high civilization? How was this culture sustained? Where are their power lines? Where are their power plants?

The engineering marvel of Egypt known as the Great Pyramid of Giza provides some answers to these questions. Its sheer size and precision have evoked amazement and wonder from people of all disciplines who for decades have studied and tried to understand what it would take to duplicate it. Moreover, it has prompted people to question and wonder about the nature of its true origins—indeed, about the true purpose for which it was originally built.

Why should this be? Hasn't it been proven that the Great Pyramid was originally a tomb? Well, it depends on whom you believe. Certainly scholars have theorized that the Great Pyramid was built to be a tomb, but their questioning and inquiries have continued for decades without ceasing, and the views of many are that the theory is not supported by evidence. In this book,

I will evaluate and present evidence that refutes the tomb theory and that shows instead that the Great Pyramid of Giza is in fact an amazing—and technically advanced—machine.

The Great Pyramid has dominated the Giza Plateau for thousands of years; and during those years it has attracted the attention of millions by its profound ability to puzzle, confound, amaze, and defy the questioning minds of generations of scholars. In the chronology of serious studies of the Great Pyramid, there has been so much wonderment, superstition, speculation, and religious awe directed toward it that it is sometimes difficult to view this structure without one of these emotions coloring one's perspective. Enormous amounts of data have been amassed about this pyramid, and much of it still requires analysis. Ultimately, researchers have had to leave the subject without completely answering all the questions. The following two quotations aptly express the dilemma faced by anyone trying to understand the true significance of the Great Pyramid. In *Secrets of the Great Pyramid*, Peter Tompkins wrote, "I have collected a mass of numerical evidence which shows that the inhabitants of the ancient world were acquainted with the rate of the precession of the equinoxes and attached a major significance to it. But in order to deal with this evidence, I would have to open an entirely new topic. I beg the indulgence of the reader in asking him to remain satisfied for the moment with the mere hint that there is yet another lesson about the level of Egyptian science to be drawn from the stark nakedness of the Great Pyramid."[1] William Fix, in *Pyramid Odyssey*, said, "Making sense of the Great Pyramid and the information encoded in it requires a fundamental re-visioning of history and the nature of man."[2]

One night in September 1977, I was engrossed in Tompkins' book, and his ideas, and those of numerous other researchers, that the Great Pyramid was more than just a tomb, resonated within me like rolling thunder. Thunder touches everything in its path, but to understand it, you have to penetrate a heavy cloak of clouds. I felt as though I was penetrating those clouds. The technologist in me was awakening to a voice that leapt from the pages and demanded attention. I became fascinated with a topic about which I had little prior information or interest. My life was changed.

Encyclopedias contained little of the data that Tompkins' book provided. His predominant focus was the Great Pyramid, and he presented theories of numerous researchers dating back to the time of Herodotus. There was a distinct presumption on the part of many that certain characteristics of the Great Pyramid did not fit the expectations one would have for a burial place. Not wanting to stray too far from the "official" theory, some assigned a dual purpose to it. Others questioned the validity of the tomb scenario and offered other ideas to supplant it. Using photographs, sketches of the inner passages and chambers of the Great Pyramid, and measurements carefully taken by nineteenth-century explorers, Tompkins presented details describing a relic from the Old Kingdom in Egypt that, when examined in the context of an undeveloped society, stood out in stark contradiction to traditional views of the building and purpose of the Great Pyramid. Moreover, the accurate detail and precision with which the Great Pyramid was built were clearly very advanced, even when compared to the efforts of modern technologists such as myself.

In my mind, Tompkins' questions were persuasive arguments for further study of the Great Pyramid, and they launched me on an individual quest to evaluate the data myself. I was driven by the question: If the Great Pyramid is not a tomb, then what is it? A large part of my background has been studying blueprints and understanding the intentions of the engineers and draftspeople who created them. Studying the drawings that showed a cross section of the Great Pyramid and reading about the astounding precision built into it, I was astonished and could find no logical resemblance to any feature one would find in a building constructed for human activity. Precise descriptions of almost every inch of the Great Pyramid revealed an accuracy and detail on such a large scale that I began to question that the Great Pyramid was used as a tomb.

I began to see the drawings of the Great Pyramid, with its numerous chambers and passageways positioned with such deliberate accuracy, as the schematics of a very large machine. I became convinced that it could not be anything else, and I set about trying to understand how this machine operated. The effort could be considered similar to what is known as the process of reverse engineering. To be successful at this, I knew that I had to find an answer for every single detail found within the Great Pyramid. I could not ignore any evidence or twist it in any way. I was determined to prepare a report that was accurate and as honest as I was capable of making it.

As a craftsman and engineer, I have worked with machines for over thirty-five years. I began to apply my specialized knowledge to the data gathered about the Great Pyramid. For instance, scholars have suggested that the pyramids were built with primitive hand tools. This is a subject I know something about. I once apprenticed in England, where I worked many hours using nothing but hand tools. Saws, drills, files, and chisels were all we were allowed to use to create precision objects. At the time, I failed to see the benefit of this toil. Why work eight hours a day bent over a piece of steel clamped in a vise when there was machinery in the area that would do the work more quickly and accurately? The result of this labor was several precision artifacts and—more important—the knowledge and appreciation of what it takes to produce something by hand. It also served to forge continuity between the craftspeople of the Old World and those of the New World. As I evaluated the opinions of Egyptologists about ancient building and machining techniques, my training told me that their theories were lacking at best—and outright wrong at worst. As I looked at the data, in fact, I began to suspect that the ancient Egyptians may have used technologically advanced tools.

Bringing applications from my work as a machinist to bear on my speculations about Egypt and its Great Pyramid, this vague suspicion became a firm belief as I pondered for innumerable hours on the methods that might have been used to produce the various artifacts. I was filled with awe and wonder at the techniques used, and at the same time I began to be aware of a greater appreciation for the technology our own society has developed. I also wondered what future archaeologists would say about some of the artifacts we may leave behind.

With the advances in manufacturing technology, my career has been a continuous educational experience that ultimately guided me into the field of laser processing of materials. During this period, I was asked to give a presentation to a local high school on career opportunities in manufacturing. In preparation for my presentation, I cut two identical cartoon characters out of stainless steel on a 5-axes YAG laser. The machine is a computer numerical controlled machine, and each character had fine detailing with a .010-inch kerf (cut width).

Having a fascination for the analytical skills necessary to determine how prehistoric societies manufactured precision artifacts, I presented one of these laser-cut figures to the class and told them that if our civilization were to be destroyed, future archaeologists may be able to determine the manufacturing tools our civilization uses just by studying that object. The surface of the cut, when studied under a microscope, would show signs of a recast layer, indicating the use of heat in its production, and the fine kerf can only be produced by that heat being focused to a very small spot. The laser, I explained to the class, is the only method that is capable of producing the features found in this object. I then described the various disciplines involved in the creation of the laser. It required physicists, optical engineers, mechanical engineers, materials and electrical engineers, and a host of craftspeople building equipment to their specifications. There were quite a few careers involved in the creation of this seemingly simple cartoon character.

To envision this laser-cut object, think of a talented artist drawing freehand with a pencil. The lines where the laser has cut through the material are as thick as a pencil line. Using the law of Occam's razor, where the simplest solution is probably the correct one, it could be assumed that a talented artist created this stainless steel character by guiding a handheld laser. I then produced the other cartoon character and placed them together so that each feature overlaid the other perfectly. Now, I told the class, because the human hand and eye are incapable of producing two objects that have complex features with such precision, the scope broadens. There were obviously other disciplines and careers that had a hand in the cutout. There had to have been some mechanical device to guide the laser along its path. There had to have been controls to turn the laser off—as it passed over the metal— and on again to punch a hole through the metal and begin once again cutting the intricate lines. We need electronics engineers, machine tool designers and builders, and computer engineers and programmers, I explained, to build the computers and write the codes that guide the machine tool and control the laser. Besides introducing the class to the hidden work opportunities that are behind the most simple artifacts, my point was to teach the students that a tool is neither created or used in isolation. What I did not tell them was that the same analytical skills and analyses that are readily accepted when applied to modern artifacts can be equally beneficial when analyzing artifacts from ancient times. The difference is that the tools that created modern artifacts are still in existence, while those that created many ancient artifacts are not.

It has been the practice of archaeologists to study the artifacts of a civilization and determine the minimum resources necessary to create them. Generally the primitive tools archaeologists uncover are sufficient to explain these artifacts. There are exceptions, however; and Egypt has an abundance of artifacts that still need to be evaluated correctly. Attempts have been made to explain some of these artifacts, but they fall short of determining how we could actually re-create the artifacts themselves. Part of this problem among academics is a persistent barrier in their beliefs which has resulted in their unwillingness to consider ancient civilizations as being advanced. It is my contention that until scholars select the methods that accurately replicate some of these artifacts under study, they will continue to underestimate ancient achievements and fail to learn their true significance.

Because so many Egyptian artifacts, including the Great Pyramid itself, cannot be explained adequately or fully by any one theory, the field of Egyptology is rife with controversy and speculation. There is no shortage of theories regarding the construction and meaning of the Great Pyramid—and the believers of a particular theory have a tendency to hold it passionately and religiously. In order to present my own view, I will address other theories and identify where they fall short. My purpose, however, is to promote cooperation between multidisciplined researchers in the quest for knowledge about our prehistoric ancestors. No single discipline is capable of analyzing and presenting the entire truth regarding the Great Pyramid. It requires experts from many different fields. And Egyptology is only one of them. The fact is that from laypeople to senior research scientists, the old theories are being rejected, and there are new views being presented by researchers with expertise in various fields. While faced with criticism and sometimes derision for their ideas, these new, often independent, theorists possess a high level of cooperation and dedication to the truth. For example, Robert Bauval, author of *The Orion Mystery*, has these qualities. His discovery of the stellar alignment of the Giza pyramids with the constellation Orion is a valuable lesson that challenges us to reconsider both the Great Pyramid's function and the ancient Egyptians' level of astronomical knowledge. Other independent researchers such as Graham Hancock, John Anthony West, and Robert Schoch have supported and energetically promoted the airing of views different from their own because they believe that each contributor to this research could bring a vital clue in our understanding of this ancient culture.

This new understanding is important to us as a species for it supplies us with a history that is deeper and richer than we previously thought. At the same time, it provides us with a guidepost to a future that combines the best of both worlds—blending the technology of the present with the technology of a past that we are only now rediscovering. And perhaps more important, this new understanding will reveal a thread of consciousness that is connected with our distant ancestors, giving us a new perspective and sense of mortality.

Chapter One

A NEW PARADIGM, A NEW ORDER

*here is excitement in the air, and the Internet is buzzing. There is* something going on in Egypt. There is intense anticipation that new discoveries and a tremendous amount of information are about to be uncovered. Why all this interest in the relics of an ancient civilization that flourished in an area of the world so re-moved from our own? Egypt has always had the power to attract and mystify. To visit Egypt and enter the massive stone edifices still standing after eons is to be drawn into a spell that has been wielding its influence for millennia. What is going on at Giza? What revelations regarding prehistory are now forthcoming?

Most information related to ancient Egypt has been in the control of Egyptologists, and it has typically been their research and discoveries that have held authority over all others. Egyptian Egyptologist Zahi Hawass, the director of the Giza Plateau, recognizes that Egypt is in possession of ar-chaeological sites that are the intellectual property of the world. At the same time, the pyramids and the Sphinx are valuable sources of income for Egypt from tourism and archaeological permits.

In recent times, expert opinions—other than Egyptologists'—have been solicited regarding the relics of Egypt. For example, during a recent explora-tion on the Giza Plateau funded by Dr. Joseph Schor, an engineer was in-vited to participate in a search for the Hall of Records, which Edgar Cayce predicted was underground near the Great Sphinx. Tom Danley, an acous-tics engineer and consultant to NASA, also conducted resonance tests inside and above the King's Chamber in the Great Pyramid. The results of his tests are remarkable and will be addressed later in the book.

The summer of 1997 was filled with reports of clandestine digging in-side the Great Pyramid. Eyewitnesses testified that fresh tunnels were being dug above the King's Chamber; and while equipping the chambers above the King's Chamber with vibration sensors, Danley discovered evidence of fresh tunnels being dug there. Who was doing the digging? Who authorized it? And what are they looking for? Local officials expressed surprise at the news and stated they knew nothing about it. Danley wrote to me in an e-mail: "Who ever was doing it was being careful to 'hide' their work, there was NO dust downstairs at the time like is mentioned at Hoagland's site[1] and the burlap bags of chips were hauled up to the next level UP and heaped against the wall, along with many water bottles and trash. I do not think that the officials knew about it either as they let us go up there with no chaperone and our inspector was very surprised when I told him I thought there was new digging going on.

"My guess would be that the officials were now simply removing the bags for safety and that's how the dust got down to the Grand hall."

In an interview on the Art Bell radio show on July 25, 1997, Danley described the tortuous path one has to take to access the upper chambers above the King's Chamber. In 1836, over a period of several months, English aristocrat Colonel William Richard Howard-Vyse created access to these chambers by blasting upward through the limestone and granite. The hole he created is more like a chimney, with rough sides enabling footholds to climb up. With the new hole, however, it was obvious the tunnelers had hauled the burlap bags of limestone chips up to the chambers above as they dug, instead of removing them completely from the pyramid. This is obviously a more difficult task and surely must have been a conscious effort to keep the digging secret. Boris Said, a documentarian and producer of the television documentary "Mystery of the Sphinx" with Charlton Heston, was being interviewed along with Danley, and he speculated that the purpose for the digging was the Egyptians' clandestine attempt to reach behind the "door" at the end of the southern shaft in the Queen's Chamber, discovered by robotics engineer Rudolph Gantenbrink in 1993. For now the issue rests, for officials are not admitting any knowledge of the new tunneling and probably will not until such time as they make a remarkable discovery.

As much as the Great Pyramid has deteriorated over the passing millennia, one would think that another hole bored into the heart of this structure would not matter much. But it seems to be creating quite an emotional stir among those who revere this edifice and who view it as an inheritance for the world rather than the personal property of the Egyptians. Even with this deterioration, though, the quality of the workmanship that went into building the pyramid is still evident, and its tremendous significance has prompted many alternative theories as to its function. What scholars and laypeople must remember is that *any* theory that purports to explain the purpose of the Great Pyramid should be mindful of each aspect of its physical existence. The material evidence found within the Great Pyramid did not just spring into existence, but was the result of a physical event, whether the event was planned or not. Thus, every single discovery, observation, and peculiarity—carelessly noted or closely scrutinized by researchers—was the result of some planned action by the pyramid builders or was the effect of a definite cause. Everything about the Great Pyramid has an answer.

The Great Pyramid is the largest, most precisely built, and most accurately aligned building ever constructed in the world. To my mind it represents the "state of the art" of the civilization that built it. (State of the art describes a condition of excellence, wherein the pursuit of any occupation and the product of that occupation is the best example of it, using the most up-to-date methods available for its completion.) There is no evidence to support the speculation that a civilization, for one brief period of time, could produce work that is so advanced it would be considered supernatural to the members of that society. We will get further in our understanding of the Great Pyramid if we follow the premise that it is an accurate reflection of the technology that was developed and used by the society that built it.

Many technologists concur that the state of the art evident in the Great Pyramid is, by modern standards, very advanced. As the technological achievements of a society advance with time, the state of the art in any particular field continually improves as new methods are implemented. The technology we enjoy today has progressed gradually over the years, and each improvement has redefined the state of the art, and with it, our lives. These improvements are not designated to just one area, and many times an improvement or a discovery in a particular science has enabled other professions to advance. As a result, a balance is maintained between the sciences; and in climbing the ladder of technological progress, one area of science may install a rung with which another may climb higher.

We are not unduly amazed when confronted with a display of our own society's technological advancements, for in viewing the end product, we are aware to some degree of the technology employed in its creation. For instance, as we stroll through our climate-controlled shopping malls, we take for granted the use of advanced machines and the high-tech methods of manufacturing and construction that make them possible. But if we were completely unaware of the techniques and machinery used to build such a complex, we would undoubtedly be stupefied as to how it came into being.

This bewilderment has affected many students of ancient cultures, particularly of Egypt, because we have been taught that the only means of construction available to the builders of the Great Pyramid were manpower, ropes, and tools of copper, stone, and wood. As researchers attempt to reconstruct—in their minds, on paper, and sometimes even physically—the achievements of these ancient, technologically "primitive" builders, they are amazed at the lack of correlation between what they see rising from the desert floor and what they "know" to be history.

Many theorists, unable to reconcile the rift between accepted theory and the fact of these magnificent structures resort to supernatural theories to explain who built the Great Pyramid and how. They surmise that:

- the Great Pyramid was built by super beings who came to Earth from another planet.
- the Great Pyramid was built through divine inspiration.
- the Great Pyramid was placed on Earth, completely intact, by the hand of God.

Still other researchers and authors of books about the Great Pyramid speculate that it embodies an ancient and lost science. They subscribe to the belief that the practitioners of this science built the pyramid as testimony to the knowledge they had developed and because they believed in the prophecy of a future cataclysmic event. David Davidson, a structural engineer from Leeds, England, theorized that the Great Pyramid is similar to a time capsule left by some fantastic civilization for the benefit of a future generation that possesses the ability to unlock its secrets and reap the benefit of the knowledge stored there.

In my view, however, the Great Pyramid reveals too much practical experience and technological knowledge on the part of its builders to suggest that they suddenly diverged onto a path of symbolism and occultism. If indeed we benefit through our study of the Great Pyramid, perhaps realizing that the ancient Egyptians were very advanced, then these more speculative theories might have some basic truth to them. As an engineer, however, I am mindful that *our* civilization's major construction projects are not financed on the collateral of some future generation thousands of years hence, but are built to serve the needs of today's society. It is likely that some of our larger construction projects would survive a world catastrophe and last for several thousand years. For example, if a disturbance around the globe returned us all to the Stone Age, the Hoover Dam, a colossal construction project of modern times, would be viewed with awe if the science and technology that were needed to build it had been lost to everyone. This dam, and others like it around the world, was not built to serve some far distant civilization, but to fill a need at the present time. Financing was provided on the basis that there would be some return on the money invested in it. It seems logical to assume, therefore, that the builders of the Great Pyramid, especially the financiers of its construction, were expecting some return on the resources they invested.

The construction date of the Great Pyramid has been speculated to be from 4,800 years to 73,000 years ago. An Arab writer, Abu Zeyd el Balkhy, estimated the oldest date from an ancient scripture. He claimed that it was built at the time when the Lyre was in the constellation of Cancer, which has been interpreted as meaning "twice 36,000 years before the *Hegira*," or around 73,000 years ago.[2] (The *Hegira*, or *hijrah*, was the flight of the prophet Mohammed from Mecca to Medina to escape persecution and is dated A.D. 622. The term hijrah also relates to the migrations of the faithful to Ethiopia, as well as those of Mohammed's followers to Medina before the fall of Mecca.[3])

The earliest date of 4,800 years was suggested after the discovery of a cartouche, or royal inscription, inside a scroll-shaped design painted on the ceiling of the top so-called construction chamber above the King's Chamber. This cartouche was supposedly the emblem of Khufu, called Cheops by the Greeks, who is said to have reigned in Egypt 4,800 years ago. Some writers have thrown doubt on the authenticity of this and other cartouches and claim that Howard-Vyse forged them while working during his 1836-37 expedition. It is suspected, from his diary—which he updated daily—that he was overly anxious to provide significant discoveries to his familial bene-factors, who had provided him with £10,000 sterling for his expedition. It is reported that members of royalty were visiting Egypt at that time, and he wanted them to view something more than just the unadorned stone. On the day he opened the chamber where the cartouches were found, Howard-Vyse made no mention of them in his diary. The following day he directed others into the pyramid to witness them. It was as if they appeared over-night.[4] Other writers insist that they are authentic. John Anthony West, dur-ing a recent telephone conversation, told me that he had recently climbed to the upper chambers and is quite convinced that the cartouches were painted on the stones at the time of the building.

The pyramids are products of a society that is known to have put a great deal of emphasis on death, the afterlife, and associated funerary trap-pings. Consequently, it is not surprising that these huge, mysterious edifices would be labeled as tombs. What else could they be? However, the Great Pyramid and its neighbors still remain a mystery to many people who have studied them. I am one among many who do not believe the tomb theory, although I recognize that there are those who see no mystery and who have satisfied themselves that this is the "true" function of the pyramids of Egypt. In all fairness, it should be stated that the theory proposed by Egyptologists has been around for quite some time and has become an unquestioned be-lief for many academics and laypeople alike. It is worthwhile to note, how-ever, that Egyptologists do not claim to know everything about the builders of the pyramids. They confess, at times, to be unsure of many aspects of the construction methods used to build them. Nevertheless, they seem unified in their belief that the pyramids were the tombs of the ancient Egyptians.

But if this is so, where are the mummies that were supposedly buried in these pyramids? According to one Egyptologist, there are not any! In 1975, during a leisurely stroll around the Giza Plateau, U.S. Egyptologist Dr. Mark Lehner told William Fix that no original burial has ever been found in any pyramid in Egypt![5] Is this a revelation to you? It certainly was to me. Still, many people identify the pyramids with the discovery of King Tutankhamen's tomb. I remember seeing an old newsreel that flipped from the Great Pyramid to the Valley of the Kings dramatically—and incorrectly—proclaiming that the valley was in the shadow of the pyramids.

A greater awareness of those who oppose these kinds of reports has tempered this example of loose reporting, and the media has increasingly raised legitimate and difficult-to-answer questions that challenge the orthodox framework of Egyptology. After all, Egyptology is not a unique branch of science that is isolated from all others. Explaining the construction and manufacturing methods of the ancient Egyptians might well require an expertise in science and engineering that many Egyptologists do not have. But even when increasingly faced with opposing views, Egyptologists gloss over the construction methods and purpose of the pyramids and many other artifacts. This is not surprising, considering that simplistic and primitive explanations do not satisfy the evidence.

In a recent interview, British Egyptologist Dr. I. E. S. Edwards lamented that there were too many pyramids in Egypt and that pyramids had received a bad name in Egyptology circles because "they have attracted too many cranks." I am not sure what he intended by that remark, but there are many people in the world today who are questioning those Egyptologists who stubbornly cling to a speculation that has little objective evidence to support it. Although Edwards does not identify specifically whom he considers to be a crank, it is generally understood by those who have an interest in the Egyptian pyramids that anyone who offers a theory opposing the official line is at risk of being labeled a crank or a "pyramidiot." To Egyptologists, a pyramidiot could be the likes of proponents of Pyramidology, the divine inspiration school whose members have included John Greaves, John Taylor, Scotland's Astronomer Royal C. Piazzi Smyth, Joseph Seiss, J. Ralston Skinner, David Davidson, and James and Adam Rutherford. They see the Great Pyramid as a bible in stone and have prepared a chronology of biblical history based on the measurements of the inner chambers and passageways of the Great Pyramid.

Those who have been skeptical of Pyramidology—but avid students of some form of alternate view—include Sir J. Norman Lockyer and the "pyramid power" people including Antoine Bovis, Karl Drbal, and G. Patrick Flanagan. Then there are the popular-selling treatments of the mystery of the pyramids from Robert Bauval, Graham Hancock, Colin Wilson, Erich von Daniken, William Fix, Kurt Mendelssohn, and Max Toth. The works of Peter Tompkins, while hinting at an esoteric, alternate viewpoint, stand apart from the other popular works in this genre by virtue of their scope, clarity of research, and presentation.

Viewpoints that differ from the official interpretation of the Great Pyramid are not uncommon. Unfortunately, new viewpoints have not always inspired respect. Nevertheless, even though opposing views regarding this ancient artifact have not had a lasting effect on the most widely believed tomb theory, there are researchers who have worked tirelessly to bring their revisionist ideas forward. In the process, they have revealed a significant amount of detail on the subject to the general public. Without their efforts, much of this information would have been forgotten or lost in some relatively arcane academic journal.

Some authors who have attempted to debunk the Egyptologists' line of thought have, it appears, unwittingly fed fuel to their academic fire by presenting highly subjective evaluations of the structures. These evaluations are sometimes based on poorly researched and one-sided data. For example, one theory has it that the pyramids were built by extraterrestrial beings as landing pads for their spacecraft.[6] If that was the case, where did the aliens initially land their craft so they could build these structures?

While Egyptologists may be stumped regarding certain aspects of the pyramids, they are justified in defending their beliefs against such speculations. Nonetheless, even though these speculations may be blind stabs in the dark, they do reflect an increasing disenchantment with the traditional interpretation of these structures. Many who oppose the tomb theory are engineers, who understand the physical requirements needed to produce large-scale engineering works, and technologists, who understand what is behind the creation of precision work.

Unfortunately, the revisionist opinions are too fragmented to have inspired any serious consideration by academia, and the Egyptologists could well use this situation as an argument for their case. One can hear them asking, "How do you expect us to consider an alternative theory for the pyramids when you cannot agree among yourselves?" Until an answer is found for the true purpose of the pyramids, and until that answer is universally accepted, the concerted voice of the Egyptologists will continue to dominate our encyclopedias and textbooks, and, subsequently, the education of our children. Until such a day, orthodoxy holds sway. After all, finding a large building containing an empty box that resembles a burial sarcophagus, does, on the surface, certainly promote the speculation that it was a tomb.

So what is all the fuss about? Why can't Graham Hancock, Robert Bauval, John Anthony West, and others who have championed new theories, accept what is "common" knowledge?[7] Why risk one's personal reputation and livelihood if there is a shred of evidence that supports the orthodox view of prehistory? I suppose it is a simple matter of having a burning desire to know and understand the truth. I have looked at the evidence, and there is no doubt in my mind that in order to understand the truth regarding the Great Pyramid, we must first discard the tomb theory and look elsewhere for answers. But first, let us look at the orthodox theory a bit more closely (see Figure 1).

There was a time when thinking the Great Pyramid was anything but a tomb may have been considered close to heresy. Nevertheless, this idea is

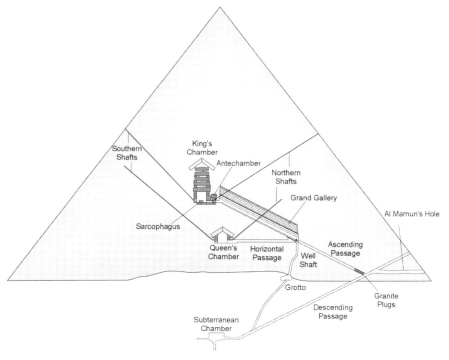

FIGURE 1. *The Great Pyramid*

not a modern fantasy of New Age seekers of truth. Other Egyptologists and researchers as far back as 1880 also have made known their doubts. Regarding the tomb theory, Piazzi Smyth wrote, "And this notion finds much favour with the Egyptologists, as a school; though facts are numerously against them, even to their own knowledge." Quoting Sir Gardner Wilkinson, an Egyptologist of that decade, Smyth continued, "Sir Gardner's gentle words, we repeat, are: 'The authority of Arab writers' (alluding to those who had described something like the dead body of a knight with a long sword and coat of mail being found in the coffer) 'is not always to be relied on; and it may be doubted whether the body of the king was really deposited in the sarcophagus (coffer) of the Great Pyramid'" [parenthetical comment within Wilkinson's quotation is Smyth's].[8]

Despite the doubts cast by Smyth and others, Egyptologists have over the years amassed as many as 20,000 publications to support their theories and they remain secure in their speculations and in the chronology they have established for the Egyptian dynasties. The lifestyles of the ancient Egyptians and of the kings, queens, and pharaohs who reigned over this society are well documented, and they are not my concern in this book. What I am interested in is just how Egyptologists propose a king or pharaoh might have directed the construction of his pyramid.

Egyptologists claim that Khufu began construction of his pyramid so it would be completed in time to accept his corpse. I should imagine that while he was considering what style of pyramid he wanted, he would have been consulting his architects and engineers to see what was feasible. He also might have been interested in knowing how long it would take to build and how much it would cost. Using today's technology, modern stonecutters have estimated that it would take at least twenty-seven years just to quarry and deliver the stone.[9] I wonder how long it would have taken Khufu's men using simple, primitive methods?

In the past, powerful leaders have erected large-scale works to satisfy their egos. India's Taj Mahal would be an example of an emperor's influence. The Mughal emperor Shah Jahan ordered it built after the death of his wife, Mumtaz Mahal, in 1631. With the concerted effort of 20,000 workers, the mausoleum building was erected in just two years, although the entire complex took twenty-two years to finish, at a cost of forty million rupees. Thus, it cannot be argued that an ancient leader could not amass the resources needed to fulfill any egotistical desires he might have about the afterlife, even if these desires and their fulfillment seem to be illogical to modern pragmatists.

There are, however, explicit engineering qualities associated with the pyramids that do not support the theory that it was a temple, a tomb, or a mausoleum. The redundancy of masonry in these structures is only one good argument against the tomb theory. More persuasive is the fact that Egyptologists woefully lack the material evidence to support it—there are no bodies! It is a widely held popular belief that Egyptian pyramids contained mummies, and that these mummies were actually discovered inside the pyramids. This is simply not true. These beliefs are only inferences that are reinforced by inaccurate documentaries that link the pyramids closely with the Valley of the Kings, where there are no pyramids, but where the mummies actually were found. In reality, the Giza Plateau and the Valley of the Kings are two vastly different sites, separated by hundreds of miles of desert. It is now becoming widely recognized by people who research the pyramid issue that of all the pyramids excavated in Egypt, *there was not one* that contained an original burial. Considering that more than eighty pyramids have been discovered in Egypt, this fact alone practically negates the tomb theory.

William Fix closely studied the subject of original burials, and he came up with some startling information regarding the absence of mummies in the pyramids: "The standard explanation for this is that every single pyramid was emptied by grave robbers in search of treasure. Grave robbery is undoubtedly one of the archaeological facts of life, and so is the later expropriation of some of the pyramids for burial purposes—a practice which at first misled archaeologists and seemed to support the tomb theory. During the Saite period (663-525 B.C.) there was an intense revival of interest in the pyramids and it became a 'fad' to use them as tombs. It is generally agreed that the coffin lid fragment found in the Third Gizeh Pyramid was stylistically a product of the Saite period, although the bones appear to be even more recent."[10]

Fix related that in 1837, sixty mummies were discovered in a large gallery under the Step Pyramid at Saqqara, fifteen miles south of Giza (see

Figure 2). It was discovered later that the mummies were interred approximately 2,400 years after the pyramid was built and not long after the gallery had been excavated beneath the existing prehistoric pyramid. Both events took place during the Saite era.

FIGURE 2. *Step Pyramid*

While we cannot assume that all individuals or groups of individuals always operate on the same principles of logic as ourselves, there has to be some firm base on which to postulate the probable actions of individuals in a given situation. It seems, therefore, sensible for Fix to write, "If only a few intact burials had been discovered, it would be easier to accept grave robbery as the fate of the others. But without so much as a single original burial, the tomb theory seems to have a large hole in it. Why would thieves seeking gold and jewels also take the corpses?"[11]

Another remarkable but little known fact concerning the alleged pyramid tombs is that while the emptiness of most of them could be blamed on grave robbers, there were some undefiled "tombs" with sealed sarcophagi that were completely empty when they were first opened. Physicist Dr. Kurt Mendelssohn wrote, "The fact that the sarcophagi in the Khufu and Khafre pyramids were found empty is easily explained as the work of intruders, but the empty sarcophagi of Sekhemket, Queen Hetepheres, and a third one in a shaft under the Step Pyramid, pose a more difficult problem. They were all left undisturbed since early antiquity. As these were burials without a corpse, we are almost driven to the conclusion that something other than a human body may have been ritually entombed."[12]

Without the presence of at least one mummy, what proof is offered to support the tomb theory? Inscriptions in the masonry of some of the pyramids have been interpreted as belonging to various dignitaries and generally are offered as the most conclusive proof of ownership of the pyramids. The presence of granite boxes that look like caskets in some of the pyramids is presented as more proof. This "proof", however, identifies only geometry and craftsmanship, not support for a theory that is highly subjective and based entirely on speculation.

The geometry and craftsmanship in the Great Pyramid have been topics of great interest and speculation for centuries. Lacking any evidence that a body was ever entombed there, but still clinging to their views, orthodox Egyptologists have been obliged to provide an explanation for the peculiar features of its passages and chambers. How do they explain the Descending Passage, Subterranean Pit, Ascending Passage, Horizontal Passage, Queen's Chamber, Grand Gallery, Antechamber, and the five superimposed chambers that overlay the King's Chamber? What explanation is given for the shafts that run from the King's and Queen's chambers to the outside? According to many Egyptologists, the entire interior complex of the Great Pyramid was the result of Khufu's, or the ancient artchitect's, indecision and symbolic reasoning. It appears that the ancient Egyptians changed their minds a lot, which resulted in some very expensive rework. I.E.S. Edwards described King Khufu as capricious in his monumental undertaking: "Externally, the Great Pyramid appears to have been completed without undergoing any significant changes in its original plan. But internally, great changes were made as construction proceeded."[13] Edwards relates that the builders dug the Descending Passage down to the Subterranean Pit with

the intention of having it serve as a burial chamber. A second chamber probably would have been added to the end of the passage that runs south from
this chamber, but, according to Edwards, the builders abandoned the entire
underground burial plan.

Having changed their minds about a subterranean burial, Edwards says,
the builders cut an opening in the ceiling of the Descending Passage and
constructed the upward-sloping Ascending Passage, the Horizontal Passage,
and then the erroneously named Queen's Chamber. However, the builders
changed their minds again. According to Edwards, the plan was abandoned
and work began on the Grand Gallery with its corbeled walls and 28-foot-
high ceiling reaching deep into the heart of the pyramid to where the granite King's Chamber is now situated. Obviously, Khufu's men were very oblig-
ing, even though they had to haul the granite from a quarry five hundred
miles away.

To many Egyptologists, therefore, the Grand Gallery is a glorious passageway to the king's final resting place, and the two chambers inside the
Great Pyramid are the result of indecision on the part of the builders or the
reigning monarch who directed its construction. All the other features of the
Great Pyramid are explained away as being either symbolic or cultic—or they
are not explained at all. For example, according to Edwards, the northern so-
called "air shaft," which pierces the mass of the Great Pyramid with gun bar-
rel precision, actually served no practical purpose and was retained only as a
symbolic gesture to the traditional downward-sloping corridors of other
tombs. He wrote, "These narrow shafts have often been referred to as air
channels, but that was not their purpose. The northern shaft was evidently a
replica in miniature of the traditional downward-sloping entrance corridor.
And so we see yet another example of an architectural element being reproduced out of its original context. It would certainly not have been retained
unless a special significance had been attributed to it." Citing a reference in
the Pyramid Texts to the constellation of Orion, in explanation of the southern air shaft, Edwards claimed, "Once every 24 hours, three stars in the constellation passed directly over the axis of the shaft. With its aid, the King
could make his ascent to their celestial region and return at will to his tomb."[14]

The explanation for the five overlaying chambers above the King's
Chamber was purely a guess at the time of their discovery. It is logical and

easy to ask, "What is this for?" But it is sometimes difficult to answer, "I don't know." So when Howard-Vyse discovered the five chambers above the King's Chamber, he speculated that they were included in the design to provide a buffer between the flat ceiling in the King's Chamber and the thousands of tons of masonry above. This guess was accepted on blind faith by others and has been repeated so often it has become ingrained. This explanation for the overlying chambers has not been questioned by Egyptologists—or any other researcher, for that matter. It may be argued that because the King's Chamber was at one time subject to a powerful force—the chamber walls show evidence of having undergone a violent repositioning—and the ceiling did not collapse, Howard-Vyse guessed correctly. However, as I will point out in more detail later in the book, this is a fallacious argument—the disturbance of the King's Chamber is attributed to an earthquake, but no other chambers suffered that fate.

The traditional tomb theory has been constantly drummed into us by documentaries, books, and movies. But despite its predominance, it continues to be questioned. On a more positive note, recently uncovered evidence has prompted some Egyptologists to revise aspects of the theory.

In 1993, Rudolph Gantenbrink, a German robotics engineer, explored the northern shaft leading from the Queen's Chamber using a specially designed robot equipped with a camera and laser pointer. Approaching a sharp bend in the shaft, Gantenbrink's robot, named Upuaut II (Upuaut means "opener of the ways") encountered an obstacle in its path. There was a length of steel pipe jammed in the passage. The pipe presumably was inserted into the shaft by early explorers with the hope of retrieving some artifacts. Not wanting to risk losing the $250,000 robot, Gantenbrink rescued it from this shaft and turned his attention to the Queen's Chamber southern shaft. During this tense mechanical expedition, Gantenbrink made a sensational discovery: At a level that is higher than the King's Chamber, his robot encountered a dead end, with what has been described as a "door" through which protruded two copper fittings. The implications of this discovery were immediately apparent, and created quite a dilemma for Egyptologists. According to their theories, Khufu changed his mind about being buried in the middle chamber in favor of a chamber higher up inside the pyramid. So if the so-called Queen's Chamber was abandoned for the higher chamber, it would not make sense for the builders to continue to create these shafts as the construction of the pyramid continued.

In order to construct these shafts, the builders had to use a tremendous amount of care. The blocks needed to be cut on an exact angle and fitted together with precision. More care would have been necessary when constructing the northern shaft because the shaft does not go straight through the pyramid. With the same angle in reference to a horizontal plane, the shaft veers to the left to bypass the Grand Gallery and then veers back again once the gallery is cleared. The blocks that were manufactured to accommodate this angle had to have been cut on a compound angle. What we have, then, are a large number of limestone blocks that are precision cut with a bottom surface that is cut on an angle (see Figure 3). Parallel to this surface a rectangular channel is cut to form the walls and ceiling of the shaft. The existence of these shafts, and the precision with which they were manufactured, cannot be explained within the framework set forth by many Egyptologists; thereby they undeniably weaken the tomb theory. I will discuss a more logical purpose for these shafts later in the book.

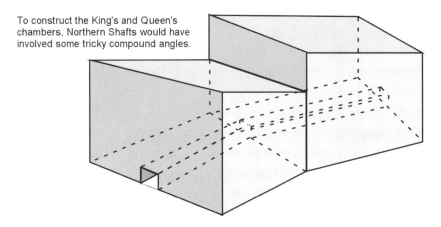

To construct the King's and Queen's chambers, Northern Shafts would have involved some tricky compound angles.

FIGURE 3. *Example of Blocks Used to Create Pyramid Shaft*

Rudolph Gantenbrink's important discovery has forced many Egyptologists to finally accept that their theories are flawed. This is an interesting development. Academic mores normally dictate that when a theory contains flaws, or unsubstantiated data that supports critical elements on which the theory is built, the entire theory must either be thrown out or revised. Instead of the tomb theory being dismissed, however, Gantenbrink himself was dismissed from the project. He discovered the "door" on March 22, 1993. A week later, he was told to pack up his robot and leave Egypt. Gantenbrink has the technology to go beyond the so-called door but, presumably because of political reasons, has been refused permission to resume his research in Egypt.

Gantenbrink, with an engineer's typical pragmatism, stated, "I take an absolute neutral position. It is a scientific process, and there is no need whatsoever to answer questions with speculation when these questions could be answered much more easily by continuing the research. Yet because of a stupid feud between what I call believers and non-believers, I am condemned as someone who is speculating. But I am not. I am just stating the facts. We have a device [ultrasonic] that would discover if there is a cavity behind the slab. It is nonsensical to make theories when we have the tools to discover the facts."[15]

Along with the recent discovery of the termination of the Queen's Chamber shaft, there is additional evidence that has been available for over a hundred years, but is seldom mentioned by Egyptologists. Cut into the bedrock about one hundred yards to the east of the Great Pyramid are features known as Trial Passages (see Figure 4). It is theorized that they were excavated to enable the workers to practice and perfect their skills before the Great Pyramid was built.

The Trial Passages are unique in that they are cut purely into the bedrock, yet they do have features that correspond with elements that are constructed, not excavated, within the Great Pyramid. A shortened version of the Descending Passage can be found as well as an Ascending Passage that is cut on the same angle as the one in the pyramid. At the juncture where the trial Descending and Ascending Passages meet there is a vertical shaft that must have fulfilled some need that the builders did not find necessary to include in the Great Pyramid. Where the trial Ascending Passage and the bottom of the trial Grand Gallery meet is an indentation that merely hints at the start of a Horizontal Passage such as that leading to the Queen's Chamber in the actual pyramid. The trial Grand Gallery displays features found in the Great Pyramid Grand Gallery, notably the steeply rising angle and the

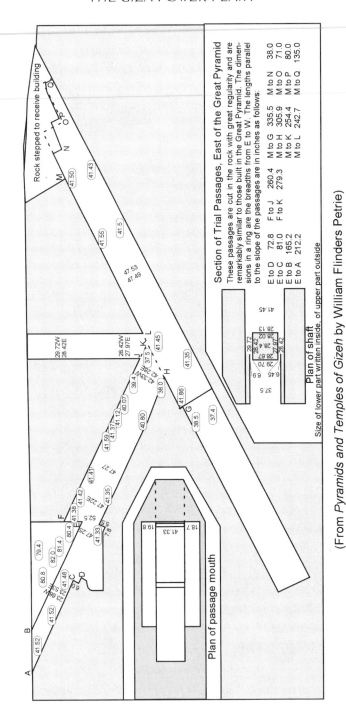

Section of Trial Passages, East of the Great Pyramid

These passages are cut in the rock with great regularity and are remarkably similar to those built in the Great Pyramid. The dimensions in a ring are the breadths from E to W. The lengths parallel to the slope of the passages are in inches as follows:

E to D   72.8   F to J   260.4   M to G   335.5   M to N   38.0
E to C   81.0   F to K   279.3   M to H   305.9   M to O   71.0
E to B   165.2   M to K   254.4   M to P   80.0
E to A   212.2   M to L   242.7   M to Q   135.0

Plan of shaft
Size of lower part written inside, of upper part outside

Plan of passage mouth

Rock stepped to receive building

(From *Pyramids and Temples of Gizeh* by William Flinders Petrie)

FIGURE 4. *Trial Passages*

18

side ramps. The dimensions and angles of all these puzzling excavations are almost exactly the same as those in the Great Pyramid.

William Flinders Petrie went further in describing these Trial Passages with tables of dimensions comparing the Trial Passages with various parts of the Great Pyramid. Table 1 is a reconstruction based on Petrie's table, using his dimensions.[16]

TABLE 1

	Trial Passages		Great Pyramid	
Passage angle	26°32'	mean difference 24'	26°27'	mean difference .4'
Passage widths	41.46	mean difference .09	41.53	mean difference .07
Passage heights	47.37	mean difference .13	47.24	mean difference .05
Ramp heights	23.6	mean difference .08	23.86	mean difference .32
Gallery widths	81.2	mean difference .6	82.42	mean difference .44
Well widths	28.63	mean difference .54	28.2	mean difference .3

The hypothesis that these passages were "trial runs" is questionable, especially when the following observations are considered:

- The passages in question were cut into the bedrock of the plateau. This would require a different technique than those used to duplicate these features inside the Great Pyramid. The excavating skills developed in digging these tunnels would, in all probability, be redundant when the builders turned to constructing the Great Pyramid.
- These "trial passages" are the only ones found on the Giza Plateau. If there had been others of lesser quality, it could be argued that the builders needed the practice, but it is evident by the close similarity between these passages and the ones in the Great Pyramid that the builders knew exactly what they were doing. They already possessed the necessary skills needed to incorporate these features inside the Great Pyramid, making such an exercise, if that is what it was, unnecessary.

It should be noted that Trial Passages were not cut for the Horizontal Passage, Queen's Chamber, Well Shaft, and Subterranean Pit—an interesting point to consider when faced with the traditional tomb theory of why the inner chambers and passages came to be. It is doubtful that both the Trial Passages and the interior chambers of the Great Pyramid were the result of indecisiveness on the part of the builders. The Subterranean Pit, which was supposedly the first chamber that was abandoned by the king, is not even included in these passages. The Queen's Chamber, purportedly the second burial chamber abandoned by the king, also is not included. The King's Chamber, the last and final burial place for the king, is nowhere to be seen in these Trial Passages.[17]

In the course of events proposed by many Egyptologists, the Queen's Chamber was built after the lower chamber was abandoned. If the builders decided to excavate the Trial Passages after they had also rejected the Queen's Chamber, we may ask why include the Descending Passage and the Ascending Passage, which must, after all, have been already built into the pyramid? The most striking detail in this investigation is that the builders went to a lot of work to excavate these Trial Passages, and, at that stage, they were placing more emphasis on the passages than on the chambers.

It is reasonable to conclude, and the Trial Passages prove, therefore, that the builders planned the Ascending Passage and the Grand Gallery before beginning construction. More than likely they also planned the King's Chamber. We can conclude, therefore, that the interior design of the Great Pyramid was conceived before the construction started, with nothing added later, be it on a whim or for any other motivation.

With a weight of evidence opposing the traditional sequence of events in the Great Pyramid, Egyptologist Mark Lehner has modified the theory to accommodate its lack of logic. In his book *The Complete Pyramids*, he wrote, "Inside Khufu's pyramid we find developments that are unique in pyramid evolution and remarkable in the entire history of architecture. Many Egyptologists have long accepted Borchardt's suggestion that the pyramid's three chambers represent two changes in plan, with the abandonment of the Subterranean Chamber, believed to be the original intended burial chamber of the king, and then the Queen's Chamber, in favour of the King's Chamber. Several clues, however, combine to make it probable that all three chambers and the entire passage system were planned together from the outset. Three chambers seem to have been the rule for Old Kingdom pyramids."[18]

Although it is probably no more than an afterthought, given to rationalize the existence of three chambers inside the Great Pyramid, Lehner's last sentence, as amply illustrated and described in his book, is not quite accurate. Djoser's Step Pyramid at Saqqara is riddled with three and one-half miles of tunnels that branch off and then converge into a central shaft at the bottom of which is a single burial chamber. The pyramid at Meidum has only one chamber. The Bent Pyramid at Dahshur has arguably two chambers and one so-called antechamber. Khafre's pyramid, which is next to Khufu's, has only two chambers. The Pyramid of Sahure at Abusir has only one chamber. Several other pyramids listed in Lehner's book also contain fewer than three chambers.[19]

By virtue of their design, the interior passageways and chambers within the Great Pyramid are difficult to explain according to the tomb theory. Orthodox explanations are strained and unconvincing, more so because Egyptologists offer differing opinions regarding the sequence of events during the Great Pyramid's construction and the intended purpose for its principal chambers. There are differences of opinion, too, between Egyptologists and professional architects regarding the establishment of its architectural attributes. In order for the tomb theory to be valid, an impossible feat must have been performed by the guardians of the Great Pyramid after the funeral procession had departed. Jammed within the lower part of the Ascending Passage are three huge blocks of granite that block the passageway that leads to the supposed burial chamber. Egyptologists propose that the blocks were originally stored in the Grand Gallery, held in position by wooden pegs inserted into slots, and then released to slide down the Ascending Passage and into position after the funeral procession had exited the pyramid. Yet architects and engineers claim that this would have been impossible and that these blocks had to have been installed as the pyramid was being built. In order for these blocks to slide down the passage, there would had to have been a half inch or more of clearance between the blocks and the passageway, whose surfaces would had to have been as smooth as glass to overcome friction.[20] The fact is that these blocks fit into the passage without any clearance on the sides; and the limestone walls, which may or may not have been smooth, would more than likely have been scoured by the harder granite as it pushed past. In addition, past these granite plugs the Ascending Passage pierces the heart of the pyramid at a 26°8' angle. Even with a clear passage—without the granite barriers—for a burial party this does not make sense, as the passage is only forty-one inches square, with barely enough room for a person to pass.

Nonetheless, in order to uphold their theory that the pyramid was indeed used as a tomb, Egyptologists must propose that the Ascending Passage was clear of obstruction. The only other way into the pyramid would have been through a small, cramped, almost vertical shaft that connects the lower Descending Passage with the Grand Gallery—certainly not a very dignified final journey for a king. So how do we reconcile the differences in opinion between Egyptologists and technologists regarding the physical realities of the theory? Obviously we are not going to re-enact the event in order to prove or disprove the theory one way or another, so the only way to settle the issue is to come up with an alternate theory that, in light of the physical evidence, makes more sense.

In this endeavor we are faced with a catch-22. The evidence cannot be explained within the parameters set by the tomb theory, so any theory that proposes that the pyramid was not a tomb is going to be immediately suspect and in all likelihood rejected out of hand. This is both good and bad. All theories should be suspect, but they should at least be objectively reviewed before being rejected. Such objectivity, in light of all the preceding arguments, can lead us to only one likely conclusion: There is precious little evidence to support the traditional tomb theory. Indeed, the evidence proves that it is altogether erroneous. Researchers who face the facts have made suggestions that the Great Pyramid must have served some other purpose. I agree. Considering the amount of effort that went into building it, and the precision of its execution, the pyramid's function must have been extremely important to its builders, more important even than serving as the final resting place for the king. So what was the function of the Great Pyramid? It is time to look at the evidence with a fresh eye and an open mind. As you will see as we progress through the book, the evidence that leads us away from the tomb theory will strongly support another theory: that the Great Pyramid was a highly sophisticated machine, with a function that was more fantastic than we have, until now, even dared consider.

Chapter Two
QUESTIONS, DISCOVERY, AND MORE QUESTIONS

round A.D. *820, Caliph Al Mamun was inspired by reports of trea-*
sures within the Great Pyramid, and directed his men to search
for an opening to the inside. Not finding such an opening, they
resorted to breaking through the hard limestone exterior by light-
ing fires against the stone, then pouring vinegar on the heated
rock. Once they were through the hard case, the softer limestone
core-masonry yielded more easily to their chisels and they proceeded to hack
out a tunnel. After blindly working for what seemed eternity, Al Mamun's
men were about to quit, when they heard a muffled sound coming from
within the pyramid. Redirecting their efforts toward the source of that sound,
they eventually connected with the Descending Passage. But their efforts
did not cease there. Finding only a long Descending Passage with a lonely
Subterranean Chamber at the end, Al Mamun turned his attention to evidence of other possible passages. The bottom side of a large granite plug in
the ceiling of the Descending Passage indicated to him that if he cut around
the granite, he would find other passages. After chiseling around three granite plugs, Al Mamun's men opened to the world the inner chambers of the
Great Pyramid. Each year, thousands of tourists follow the path that Al
Mamun carved into this structure.

After Al Mamun's fruitless and disappointing search for treasures, there
was little attention paid to the edifice, with the exception of using it as a
quarry. Bats took over the inner passages and chambers, and suspicion took
over the minds of the local inhabitants. Without modern illumination, few
would dare to go inside, especially at noon and sunset, when a naked woman
with large teeth who seduced people and drove them insane reportedly
haunted the pyramid. Rabbi Benjamin ben Jonah of Navarre reported that

"the Pyramids which are seen here are constructed by witchcraft."[1]

After Europeans started to travel to Egypt, information regarding the wonders found there began to find an audience in Western civilization. Fueled by their own curiosity and this intense interest at home, European explorers in the area were quite energetic in studying, searching, and noting just about anything, no matter how seemingly insignificant, pertaining to the Great Pyramid. As one researcher followed another, more knowledge of this pyramid was uncovered and revealed to the world.

John Greaves, a British mathematics teacher and astronomer, visited the Great Pyramid in 1638.[2] He made studies with which he hoped to find information establishing the dimensions of the planet. During his explorations, Greaves discovered what was to be known as the Well Shaft. The Well Shaft is located at the bottom of the Grand Gallery through an opening in the west wall and is approximately three feet wide. The notches cut into the sides enabled Greaves to lower himself into the bat-infested bowels of the pyramid.

Climbing down, Greaves reached a level that was sixty feet below the level of the Grand Gallery. Here he came across a small round chamber cut into the limestone bedrock. Beyond this small cavern, and deeper still, the shaft continued downward. Not knowing what lay beneath him or whether a bottomless pit might swallow him up, Greaves dropped a lighted flare down the hole. He noted that the flare continued to flicker from the depths and assumed that the shaft terminated at that point. Deciding that he had crawled around enough for one day, Greaves made his way out into the fresh air, leaving the stifling shaft to its resident bats.

This discovery left Greaves extremely puzzled, for the Well Shaft did not seem to serve any purpose. The cavern, which is now known as the Grotto, was equally perplexing. It seemed pointless to Greaves to dig a shaft to nowhere and to enlarge a part of the shaft into a grotto. This perplexity affected later explorers as well.

In 1765, Nathaniel Davison, while vacationing in Egypt, was able to carefully explore the Great Pyramid. Going farther than Greaves, Davison was lowered by rope another one hundred feet below the level of the Grotto. Here he encountered a blockage in the shaft. Why anyone would go to the trouble of digging a shaft, with no apparent purpose or destination, almost two hundred feet into the heart of the pyramid, was a mystery to Davison.

A part of this mystery was to be solved when G. B. Caviglia, the Italian captain of a Maltese ship flying the British ensign, quit his maritime occupation and undertook the task of exploring the Great Pyramid. Caviglia was determined to shed some light on the mystery of the Well, and, after being lowered past the level of the Grotto by some Egyptian helpers, he attempted to clear the blockage that Davison had encountered before him. The blockage appeared to be just loose sand and rock, so Caviglia filled baskets with the debris and had the helpers raise the baskets up and out of the shaft. He could not persuade them to work for long, though, for the air became so foul with clouds of dust and the stench of bat dung that the men were about to suffocate. Caviglia burned chunks of sulfur in an attempt to purify the air, but this ploy did not impress his helpers, who refused to continue working.

Still determined to find some reason for the shaft, Caviglia decided to clear the Descending Passage down to the Subterranean Pit. Al Mamun's men had used this passage as a dumping ground when they were cutting around the plugs that filled the Ascending Passage. With his helpers back on the job and carrying the chippings out of the pyramid, Caviglia slowly and painfully inched his way downward.

His extreme discomfort was eventually rewarded when he discovered a low doorway on the west side of the passage. Through this doorway, a hole bored upward into the heart of the pyramid. The smell of sulfur was evident inside the doorway, and Caviglia deduced that perhaps this smell was from the sulfur he had previously burned. Digging upward, Caviglia and his workers, with limestone chips and dust showering down on them, finally broke through into the Well Shaft, thereby completing the connection between the lower parts of the Descending Passage and the Grand Gallery.

Caviglia, like Greaves and Davison before him, was still faced with the same questions: Who dug the Well, when was the Well dug, and why? Another aspect of this same mystery, which further increased his perplexity, is that from the junction of the Grand Gallery and the Horizontal Passage down to the level of the bedrock, it appears that the Well Shaft actually had been included in the original plans for the construction of the pyramid. From the level of the Grand Gallery down to the bedrock, the walls of the Well Shaft are symmetrical in their construction, and, although they do not have the precise, fine finish that is evident in other passages and chambers, their features do not resemble those of a tunnel cut as an afterthought through solid masonry, such as the forced passage dug by Al Mamun's men.

It has been speculated that this shaft was dug by grave robbers who broke into the pyramid to strip it of its treasures. This theory has been refuted by some who have debated whether or not a band of thieves would have the knowledge, perception, or sheer luck to dig a blind passage with such accuracy that it would eventually meet with the Grand Gallery, which is only a few feet wide.

In contradiction to the grave robber theory, David Davidson, a structural engineer from Leeds, in the north of England, developed a scenario of prehistoric events that, in his mind, met the demands of logic and common sense, and at the same time explained the existence of the Well Shaft.

Davidson, after spending several months studying the pyramid, felt that the Great Pyramid was not originally intended for the use of the people who built it or their king; rather, it was designed to be used as a "time capsule" in which knowledge would be preserved for the benefit of a future civilization. In a professional capacity, he also maintained that the plugs inside the Ascending Passage were positioned as the level of the pyramid grew higher. According to Davidson, to have slid them down the passage without them jamming in the process would have been an unlikely feat, as the clearance at the sides of the passage would not have been sufficient to allow them to pass freely.

Davidson's scenario was set shortly after the Great Pyramid was built, or not many generations after—before knowledge of the design of the interior was lost or forgotten. He theorized that following a violent earthquake, or some other equally devastating occurrence, the guardians of the Great Pyramid noticed some subsidence effects of the structure on the outside. Fearing that the King's Chamber also might have suffered from the disturbance, they decided to enter the pyramid to investigate. To do this, they started to dig upward near the bottom of the Descending Passage. Davidson explained that instead of taking a possible shorter route, such as taken by Al Mamun at a later date, the guardians chose their route so that they could inspect two large fissures in the bedrock of the Descending Passage.

Although these fissures can be seen in the Descending Passage, what knowledge did the guardians of the pyramid have that assured them that the unseen portion of the fissures followed a predictable direction? Would they have been able to plot the course of their tunnel with assurance that, as they bored through the limestone bedrock, they would cross the same fissures at two points? Without this assurance, would such a difficult and time-consuming project be undertaken?

Knowing that the Great Pyramid's thirteen-acre base was surveyed using modern instruments and found to be level within 7/8 inch—an astounding accuracy by modern standards—it does not make much sense to suggest that the guardians would be concerned about any subsidence. Moreover, it would be interesting to know what extent of damage the guardians thought these fissures would have on the King's Chamber, which, after all, was their primary concern. The King's Chamber is located 175 feet above the ground level. There were no disturbances noted in the Queen's Chamber or the Horizontal Passage. In 1881, Sir William Flinders Petrie surveyed the Descending Passage, where the fissures were noticed, and discovered that it was remarkably accurate. He found the passage to have an error of only .020 inch over the 150-foot length of the constructed portion; and the entire length of the passage, both constructed and excavated, was within a minuscule quarter inch over 350 feet.[3] This hard data brings into question the widely held theory that an earthquake was the cause of the disturbance in the King's Chamber. It seems obvious that those who knew the interior design of the pyramid dug the shaft. Considering the amount of work involved in digging it, the Well Shaft that connects the Descending Passage with the lower portion of the Grand Gallery must have been a part of the original design of the pyramid and served a specific purpose for its builders.

Nevertheless, Davidson's theory is given the benefit of the doubt by Peter Tompkins, who said, "There is nothing inherently illogical about this version of events. It would have been no easy job to tunnel upwards through the solid rock and various courses of masonry—altogether hundreds of tons of material would have had to be chipped away and taken out of the pyramid up the descending passage—but it would not have been impossible."[4]

It is absurd to propose that this feature of the Great Pyramid exists through the efforts of tomb robbers who were digging blindly on the chance that they might discover a burial chamber. The physical demands are monumental. Digging upward, the workers would be contending with a small, cramped, almost vertical tunnel in which they would need physical support, light, and oxygen. As they hacked away at the face of their bore, the air would be heavily laden with limestone dust, and fragments would be falling on them and the workers below. The sheer human effort would have been daunting as hundreds of tons of chips were wrestled up the Descending Passage and removed from the pyramid.

Compare the digging of the Well Shaft with Al Mamun's hole. Al Mamun's men, as the story goes, were ready to give it up after digging only one hundred feet. Their patience had reached its limit. It could be asked, therefore, how far this band of treasure seekers would have gone if the task before them had been the Well Shaft. Tompkins went on to say, "What militates against this theory is the observations of Maragioglio and Rinaldi that the walls of the Well Shaft upward from the Grotto are built and lined with regular blocks of limestone, apparently as a feature of the original structure."[5]

In their book, *L'Architettura delle Piramidi Menefite*, Celeste Maragioglio and Vito Rinaldi proposed that the Well Shaft was dug to provide air to workers in the pyramid. Egyptologist and world authority on the pyramids, I.E.S. Edwards, agreed that it may have been used for this purpose, but claimed that within the context of the tomb theory, it would not have been necessary as the Ascending Passage was open until the funeral procession had left and the granite blocks had been lowered into position at the mouth of the passage. Edwards claimed the Well Shaft was an escape route for those who facilitated the lowering of the granite blocks.[6] However, that theory—to me—presents a convoluted nonsensical way of going about things, and is barely worth the time we could spend arguing about it. I find the observations of Petrie more agreeable and likely. He cast doubt on the intentions of the builders, as interpreted by Maragioglio, Rinaldi, and Edwards:

> The shaft, or "well," leading from the N. end of the gallery down to the subterranean parts, was either not contemplated at first, or else was forgotten in the course of building; the proof of this is that it has been cut through the masonry after the courses were completed. On examining the shaft, it is found to be irregularly tortuous through the masonry, and without any arrangement of the blocks to suit it; while in more than one place a corner of a block may be seen left in the

irregular curved side of the shaft, all the rest of the block having dis-
appeared in cutting the shaft. This is a conclusive point since it would
never have been so built at first. A similar feature is at the mouth of
the passage, in the gallery. Here the sides of the mouth are very well
cut, quite as good work as the dressing of the gallery walls; but on the
S. side there is a vertical joint in the gallery side, only 5.3 inches from
the mouth. Now, great care is always taken in the Pyramid to put
large stones at a corner, and it is quite inconceivable that a Pyramid
builder would put a mere slip 5.3 thick beside the opening to a pas-
sage. It evidently shows that the passage mouth was cut out after the
building was finished in that part. It is clear then, that the whole of
this shaft is an additional feature to the first plan.[7]

Petrie also noted a large block of granite at the level of the Grotto. The
block was positioned as though it had been pushed aside from the vertical
section of the shaft. What was this block of granite doing down a shaft, which,
in many minds, had nothing to do with the original design of the pyramid?

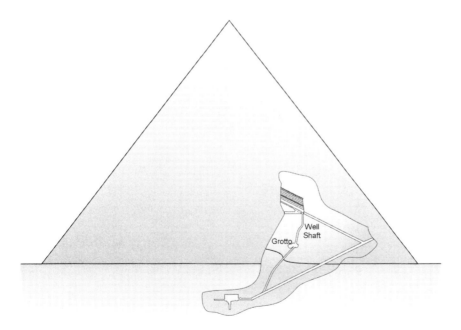

Figure 5. *Well Shaft*

After tunneling through virgin rock and reaching the level of the Grand Gallery, why would anybody decide to drop a block of granite down the Well? This would take considerable effort. And where did they get the granite? There has to be a reason for it being there (see Figure 5).

Maragioglio, Rinaldi, Edward, and Petrie's observations can be reconciled by proposing that the constructed portion of the Well Shaft was originally smaller than it is now and was enlarged to allow passage into the Grand Gallery for inspecting the damage in the King's Chamber. If this is true, then perhaps the Well Shaft was designed not for human access, but for something else. Whatever this purpose was, perhaps it called for the inclusion of a large granite block within the passage to serve a specific function. These are questions I will answer later in this book.

If the Well Shaft had been dug by grave robbers, they would have needed to know the internal arrangement of the pyramid, and be sufficiently inspired by what was contained within it to undertake the project. As for the inspection theory, we might ask, "How were the guardians able to discern any subsidence on the outside of the pyramid?" With an error of only 7/8-inch over the entire thirteen-acre base, over a distance of one foot the amount of error would be only .001 inch—less than half the thickness of a human hair! To detect the 7/8-inch error in the base of the Great Pyramid, even if the base were perfectly flat originally, the ancient guardians would had to have been in possession of some remarkably advanced measuring equipment.

Assuming that they had the equipment to measure this minute variation in the pyramid's levelness, would such an insignificant deviation from accuracy warrant the penetration of the pyramid in the manner theorized? If the guardians were concerned about the fate of the internal passages and chambers, they could have put their minds at rest by closely checking the Descending Passage. As I mentioned earlier, this passage was scrutinized by Petrie who reported, "The average error of straightness in the built part of the passage is only 1/50 inch, an amazingly minute amount in a length of 150 feet. Including the whole passage the error is under 1/4 inch in the sides and 3/10 on the roof in the whole length of 350 feet, partly built, partly cut into the rock."[8]

Nonetheless, evidence indicates that an inspection of the inside chambers of the Great Pyramid was conducted in antiquity and repairs were made. For example, plaster was daubed on the cracks of the ceiling beams above the King's Chamber. But what triggered the guardians' concern for this chamber? If they were to initiate a close inspection of the internal passages and chambers of the Great Pyramid following an earthquake, wouldn't the Descending Passage satisfy their curiosity and put their minds at ease? It would seem that the penetration of the internal chambers of the pyramid was prompted not by the detection of minute subsidence on the outside, but by other observations of the pyramid's form and function.

At the time of Al Mamun's exploration of the Great Pyramid, many untold mysteries remained hidden from his searching eyes. It was not until 1765 that Nathaniel Davison made a discovery that initiated continuing explorations and further compounded the mystery of this enigmatic structure.

Close to the King's Chamber, on the Great Step, a curious echo coming from the ceiling of the Grand Gallery caught Davison's attention. Assisted by the illumination of an elevated candle, Davison scrutinized the gallery ceiling and vaguely discerned an opening near the top. Taking his life into his hands, he erected a precarious collection of ladders and gingerly proceeded to climb to the top.

To his delight, Davison discovered that there was indeed an opening. His pleasure, though, soon turned to disgust as he wedged himself inside the hole and found himself immersed in pungent mounds of bat dung. Using a handkerchief to protect his offended nose, he forced himself into the fetid passage and struggled along in this manner for twenty-five feet until he came across a large chamber not quite high enough to stand in. Once inside this chamber, Davison cleared away bat dung and uncovered nine enormous granite beams measuring up to twenty-seven feet long and weighing up to seventy tons each. This monolithic ensemble formed the ceiling of the King's Chamber. Unlike the bottom and sides of the beams, though, the tops of them were rough-hewn with no pretension to straightness or accuracy. Davison also noticed that the ceiling of the chamber he had discovered was constructed with a similar row of granite beams. Davison could make little sense of these features, and his only satisfaction in his discovery was to carve his name on the wall and have the chamber named after himself.

Additional exploration of Davison's Chamber came in 1836 when Colonel Howard-Vyse, with the help of civil engineer John Perring, made extensive explorations of the pyramid complex at Giza. In Davison's Chamber, Howard-Vyse noticed a crack between the beams of the ceiling. He perceived the existence of yet another chamber above the one he was occupying. Without obstruction, he was able to push a three-foot-long reed into the crack. Howard-Vyse and his helpers then made an attempt to cut through the granite to find out if there was indeed another chamber above. Finding out in short order that their hammers and hardened steel chisels were no match for the red granite, they resorted to using gunpowder. A local worker, his senses dulled by a supply of alcohol and hashish, set the charges and blasted away the rock until another chamber was revealed.

This chamber held a mystery for the early explorers who entered its confines, a mystery that has baffled people for decades. The chamber was coated with a layer of black dust, which, upon analysis, turned out to be exuviae, or the cast-off shells and skins of insects. There were no living insects found in the Great Pyramid, which made this discovery even more mysterious. What prompted hordes of insects to single out this one sealed chamber and shed their skins? It is a mystery that has never been satisfactorily explained. In fact, there has not been any attempt to explain it, and because there is no logical answer that fits in with any previously proposed theory, no one has given it much attention.

As in Davison's Chamber, a ceiling of monolithic granite beams spanned this new chamber, indicating to Howard-Vyse the possible existence of yet another chamber above. Blasting their way upward for three and a half months, to a height of forty feet, they discovered three more chambers, making a total of five. The topmost chamber had a gabled ceiling made of giant limestone blocks. Howard-Vyse surmised that the reason for the five superimposed chambers was to relieve the flat ceiling of the King's Chamber of the weight of thousands of tons of masonry above. Although most researchers who have followed have generally accepted this speculation, there are construction considerations that cast doubt on this theory and prove it to be incorrect.

What Howard-Vyse and others have not considered is that there is a more efficient and less complicated technique in chamber construction elsewhere inside the Great Pyramid. The structural design of the Queen's Chamber negates the argument that the chambers overlaying the King's Chamber were designed to allow a flat ceiling. The load of masonry bearing down on the Queen's Chamber, which itself is situated below the King's Chamber, is greater than that above the King's Chamber. Yet the Queen's Chamber has a gabled ceiling, not a flat one. If a flat ceiling had been required for the Queen's Chamber, it would have been quite safe to span this room with one layer of beams similar to those above the King's Chamber. Both the King's Chamber and Queen's Chamber employ huge gabled blocks of limestone that transfer the pressure of the above masonry to the outside of the walls. The fact is that a ceiling similar to the one in the King's Chamber could have been used in the Queen's Chamber, and, as with the beams above the King's Chamber, the beams would be holding up nothing more than their own weight (see Figure 6).

When the builders of the Great Pyramid constructed the King's Cham-

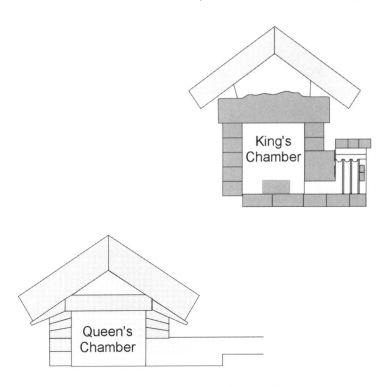

FIGURE 6. *Queen's Chamber and King's Chamber with Flat Ceiling*

ber, they were obviously aware of a simpler method of creating a flat ceiling. The design of the King's Chamber complex, therefore, must have been prompted by other considerations. What were these considerations? Why are there five superimposed layers of monolithic seventy-ton granite beams? Imagine the sheer will and energy that went into bringing just one of the forty-three granite blocks a distance of five hundred miles and raising it 175 feet in the air! There must have been a far greater purpose for investing so much time and energy—and there is, if we understand the Great Pyramid as a machine. But before I make my case, we must examine more of the evidence and the orthodox theories proposed to explain it.

With the discovery of the five superimposed "chambers of construction" above the King's Chamber, the granite complex located at the heart of the Great Pyramid became a mystery and source of consternation. The reason the builders changed from limestone to extremely hard granite could be adequately explained if we consider egos, whims, and religious beliefs to be a driving force in the decision-making process of the ancient builders. However, there are other notable peculiarities concerning the state of some of this granite that do not fit into a logical pattern.

If we focus our attention on the granite-lined Antechamber, we perceive a marked discrepancy between the craftsmanship displayed there and the meticulous care maintained throughout the rest of the Great Pyramid. Other researchers have noticed this anomaly as well. Petrie was astounded at the gross negligence and inferior workmanship he saw in this chamber, and wrote, "In the details of the walls, the rough and coarse workmanship is astonishing, in comparison with the exquisite masonry of the casing and entrance of the pyramid; and the variation in the measures taken shows how badly pyramid masons would work."[9]

Perhaps researchers are being somewhat hard on the masons who worked on the granite complex. After all, granite is an extremely hard material with which to work, and once the pyramid was closed up, who would see it? As Petrie noted, the casing stones and entrance passage of the pyramid were tooled to a remarkable closeness and fine finish. But then, that work is in a location where any deviation from the precision of which the workers were capable would be seen by all who passed by, reflecting badly on them. It is understandable that a worker, or group of workers, would

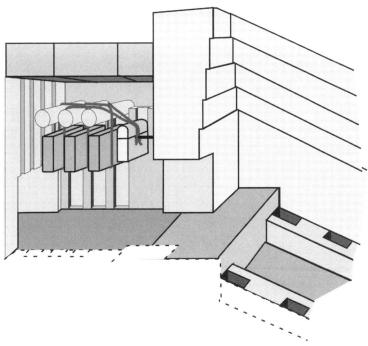

FIGURE 7. *Borchardt's Theory on the Antechamber*

produce the best work where it would be noticed and, perhaps, be less critical of it where it is hidden. But, as I said earlier, *everything* in the Great Pyramid has an explanation, and I believe my theory will adequately account for
this variation in workmanship.

Even if a reasonable explanation for the less-than-perfect work in the
Antechamber is accepted, the characteristics of the chamber still initiate a
great deal of thought. Egyptologist Ludwig Borchardt theorized that this
small chamber contained a mechanism that closed off the access to the King's
Chamber after the funeral procession made its exit. This was supposedly
accomplished by lowering granite slabs, suspended by ropes down the slots
located in the east and west walls of the chamber (see Figure 7).[10]

In explaining the "disappearance" of these slabs, Egyptologists claim
they were removed by grave robbers. Once again, the poor old grave robber
gets the blame! Of course, after digging the tortuous Well Shaft with such
remarkable precision, and luckily finding himself in the Grand Gallery, nothing would be impossible for this ingenious thief. Still, what would a grave

robber want with chunks of masonry? What possible value would granite chips have to someone with easy access to an abundance of masonry available on the outside of the pyramid?

In addition, how were those granite slabs removed if they were installed in the slots in the chamber walls? They would had to have been chipped away or lifted up in the slots until they were free. Although it has been conjectured that the slabs were lowered into position by ropes, or even the release of sand, it is unlikely that the means to reverse this process would have been left readily available to those with the intent to plunder. The alternative, therefore, would have been to chip away the granite barriers. But yet again, the remarkable grave robber is credited with accomplishing a feat found impossible by others. The record shows that on two separate occasions granite barriers have proved too hard to be cut away with simple hand tools. Al Mamun had to cut around the granite plugs in the Ascending Passage, and Howard-Vyse resorted to detonating gunpowder charges, for even in his modern time (as compared to the time of Al Mamun), an effective cutting tool for this granite was not available.

It is possible that, if the granite slabs were thin enough, they could have been broken by sheer pressure, but if the slots inside the Antechamber are any indication of the slabs' thickness, they would have been 21.6 inches thick, which is quite a hefty piece of stone. A conveniently located cleavage plane on each of the granite slabs might have enabled the grave robbers to break through, if they were able to apply the necessary pressure to the granite. Nevertheless, the very idea that granite slabs were once in place in the Antechamber is pure speculation. However—discounting the previous argument regarding the possibility of grave robbers breaking through to the King's Chamber—we can still respect the argument laid down by Piazzi Smyth:

These three grand, flat, vertical grooves, then, on either side of the narrow ante-chamber, have been pronounced long since by Egyptologists to be part of a vertical sliding portcullis system for the defense of the door of the King's Chamber. There are no blocks now to slide up and down these grooves, nor have such things ever been seen there: but the gentlemen point triumphantly to a fourth groove, of different order, existing to the north of all the others, indeed near the north-beginning of the ante-chamber; and with its portcullis block, they say, still suspended, and ready for work.

That alleged block, however, contains many peculiarities which modern Egyptologists have never explained ; and as it was first carefully described by Professor Greaves under the appellation of the "granite leaf" (from the so-called "leaf" or "slat" or sliding door over the water-way of a lock-gate in an English navigation canal), we had better keep to that name.

Its groove, instead of being 21.6 inches broad, like the others is only 17.1 broad ; and in place of being like them cut down to, and even several inches into, the floor, terminates 43.7 inches above that basal plane; so that the leaf's block, or rather blocks—for it is in two pieces, one above the other—stand on solid stone of the walls on either side, and could not be immediately lowered to act as a portcullis, though an Emperor should desire it. Nor would they make a good portcullis if they were to be forcibly pushed, or chiseled down in their vertical plane, seeing that there are 21 inches free end space between the leaf and the north entering wall and doorway, where a man may worm himself in, in front of that face of it ; and 57 inches above the leaf's utmost top, where several men might clamber over ; and where I myself sat on a ladder, day after day, with lamps and measuring-rods, but in respectful silence and generally in absolute solitude, thinking over what it might mean.[11]

Smyth's meditation on the Antechamber, its granite leaf, and the slots in the east and west walls left him resigned to a conclusion that does not seem to have been improved upon after the passing of a complete century. Writing in 1880, Smyth sagely concluded that "the granite leaf is, therefore, even by the few data already given, a something which needs a vast deal more than a simple portcullis notion to explain it. And so do likewise the three broader empty pairs of grooves to the south of it, remarkable with their semi-cylindrical hollows on the west side of the chamber."[12]

How correct, then, are those theories that explain the existence of the Antechamber? Not very, for none adequately explains the indisputable amount of work that went into making it more than just a simple room with four walls, floor, ceiling, and two passageways. There must have been a reason for the additional effort expended in cutting the four slots in the chamber walls and installing a granite slab in an immovable position. As we will soon discover, one explanation that does explain the Antechamber is that it had a mechanical function. The evidence is plainly clear if one knows how to read it.

For example, the presence of half-round hollows in the top surface of the granite wainscot definitely suggests that cylindrical objects were at one time suspended across the width of the Antechamber and that these may have been receptacles for bearings, or were the bearing surfaces themselves. Again, Piazzi Smyth took careful notes:

> . . . Little indeed is the ante-chamber, when it measures only 65.2 inches in utmost breadth from east to west, 116.3 long from north to south, and 149.4 high ; but it has a sort of granite wainscot on either side of it, full of detail ; and was to me so complicated and troublesome a matter as to occupy three entire days in measuring.
>
> On the east side, this wainscot is only 103.1 inches high, and is flat and level on the top ; but on the west side it is 111.8 inches high, and has three semi-cylindrical cross hollows of nine inch radius, cut down into it, and also back through its whole thickness of 8.5 to 11.7 inches to the wall. Each of those semi-cylindrical hollows stands over a broad, shallow, vertical, flat groove 21.6 inches wide, 3.2 inches deep, running from top to bottom of the wainscot, leaving a pilaster-like separation between them. The greater part of the said pilasters has indeed long since been hammered away, but their fractured places are easily traced ; and with this allowance to researchers in the present day, the groove and pilaster part of the arrangement is precisely repeated on the east side, within its lower compass of height.[13]

Is it conceivable that the pyramid builders went to such a great amount of trouble to cut this granite for a one-time operation? If this chamber was designed to be a closing mechanism, and it was to be activated only one time, it would not have been necessary to include such a complicated design and to cut that design out of such hard and durable material. Still, if we try our hardest to give this theory its due, we would have to admit that there are cases today where a tool or machine may be "over-designed" to do the work for which it was built. But we must be aware that for the Antechamber to find a parallel in modern industry, we would have to allow for an equivalent situation, such as an expensive die being built from the finest quality tool-steel, even though it would be used to produce only one part. The theory strains under the weight of such an unlikelihood.

Under the Egyptologists' present theories, the Antechamber is, indeed, a perplexing and contradictory inclusion in the design and building of the Great Pyramid. But there is a reasonable answer to this mysterious and puzzling feature of the Great Pyramid—it is a mechanical and technological answer that so far has not attracted any consideration.

Perhaps the clue to answering this question lies inside the King's Chamber, or perhaps above the King's Chamber in the superimposed "construction chambers." A thorough investigation of the King's Chamber by Petrie revealed that the chamber had, at one time, been subject to a violent disturbance, which had shaken it so badly that the entire chamber was caused to expand approximately one inch! The granite beams on the south end of the chamber were wrenched loose and cracked through, indicating a powerful destructive force. Petrie attributed this disturbance to an earthquake, which has been the general assumption since. In Petrie's words, "All these motions are yet but small—only a matter of an inch or two—but enough to wreck the theoretical strength and stability of these chambers, and to make their downfall a mere question of time and earthquakes."[14]

Here again something does not seem to add up. It has been accepted that an earthquake could be the only disturbing force affecting the King's Chamber, and yet we could bring the same argument into play here that we used to refute the speculation that the ancient guardians noticed subsidence on the outside of the pyramid. If an earthquake had disturbed the King's Chamber to the extent that several giant granite beams were cracked and the entire chamber was expanded a whole inch, wouldn't it be reasonable to find similar disturbances elsewhere in the Great Pyramid? The King's Chamber is located 175 feet above ground level, and yet on the lower levels of construction, no similar disturbances have been noted. On the contrary! These areas show remarkable precision—a precision that has astounded those who have researched and measured the Great Pyramid and many who have subsequently studied those findings.

The King's Chamber, it appears, shows a greater amount of discrepancy than the entire thirteen-acre base of the pyramid! Why would an earthquake seek out one lonely chamber in a giant complex of masonry, passages, and chambers? The Queen's Chamber seems to have been unaffected by this catastrophic event. The Descending Passage—as mentioned earlier—is remarkably precise. No unusual disturbances were noted inside the Grand Gallery; even the Antechamber does not show the extent of damage suffered by the King's Chamber. More important, it is the specific characteristics of the disturbance that give rise to serious misgivings about the earthquake theory. Something caused the King's Chamber to expand! This small granite chamber, surrounded by a giant mass of limestone masonry, apparently pushed against that encompassing weight to the extent that the walls were moved outward from their original position. Petrie explained the damage:

> The King's Chamber was more completely measured than any other part of the pyramid; the distances of the walls apart, their verticality in each corner, the course heights, and the levels, were completely observed. On every side the joints of the stones have separated, and the whole chamber is shaken larger. By examining the joints all round the second course, the sum of the estimated openings is 3 joints opened on N. side, total = .19; 1 joint on E. = .14; 5 joints opened on S. = .41 ; 2 joints on W. = .38. And these quantities must be deducted from the measure, in order to get the true original lengths of the chamber. I also observed, in measuring the top near the W., that the width from N. to S. is lengthened .3 by a crack at the S. side.[15]

It would be interesting to find out what pressure would be needed to move the walls and affect the chamber this way, especially taking into consideration that all the spaces above the King's Chamber also were affected by the disturbance. Petrie continues with his observations:

> These openings or cracks are but the milder signs of the great injury that the whole chamber had sustained, probably by an earthquake,

when every roof beam was broken across near the south side; and since which the whole of the granite ceiling (weighing some 400 tons) is upheld solely by sticking and thrusting. Not only has this wreck overtaken the chamber itself, but in every one of the spaces above it are the massive roof beams either cracked across or torn out of the wall, more or less, at the south side; and the great eastern and western walls of limestone, between and independent of which, the whole of these construction chambers are built, have sunk bodily.[16]

Several facts support the speculation that the guardians of the Great Pyramid were aware of the damage suffered by the King's Chamber. The hole Davison discovered, which in turn led to the discovery of the superimposed chambers above the King's Chamber, can be explained by surmising that the guardians carried out a close inspection of the damage affecting the upper levels of the granite complex. The guardians, after satisfying themselves that no further attention was required, terminated their inspection at that point.

Another point that satisfies this theory is an attempt to rectify, to some extent, some of the damage in the King's Chamber. Again Petrie related some pertinent facts concerning the ancient guardians' inspection: "The roofing beams are not of 'polished granite,' as they have been described; on the contrary, they have rough-dressed surfaces, very fair and true so far as they go, but without any pretence to polish. Round the S.E. corner, for about five feet on either side, the joint is all daubed up with cement laid on by fingers. The crack across the Eastern roof-beam has been also daubed with cement, looking therefore, as if it had cracked *before* the chamber was finished. At the S.W. corner, plaster is freely spread over the granite, covering about a square foot altogether."[17]

The cracks were evidently unacceptable to the guardians, and required the addition of a layer of plaster. The question that arises is, what purpose does a thin layer of plaster serve? It is doubtful that it would lend any structural improvement to the granite complex, for what could a thin layer of cement do to prevent one of the forty-five- or seventy-ton granite beams above the King's Chamber from collapsing? Would the cement have been added to the cracked beams for some other purpose? Whatever the answer, it was evidently important enough that the Great Pyramid was entered, at the expense of a great deal of time and trouble, to make repairs.

If we could talk with those who entered the Great Pyramid to make these repairs, what would they tell us regarding the nature of the disturbance in the King's Chamber? And how would they explain the fact that the chamber had expanded? Would they confirm the earthquake theory? Or would they gently inform us that if an earthquake had indeed shaken the chamber, it would, in all probability, have collapsed? Furthermore, how would they explain their even knowing that such a disturbance had affected the well-insulated King's Chamber? Could they convince us that the minute variation from true level found in the base of the pyramid justified their effort in inspecting the small chamber at its heart? Let us face facts—the guardians would have a lot of explaining to do.

There are more questions raised by the King's Chamber. At first glance, it appears to be just a room made from red granite. As we look closer, though, it poses more mysteries than the rest of the chambers and passages of the Great Pyramid combined. While poking around the pyramid, John Greaves partially uncovered one of these mysteries.

Greaves was puzzled by the many features of the Great Pyramid that

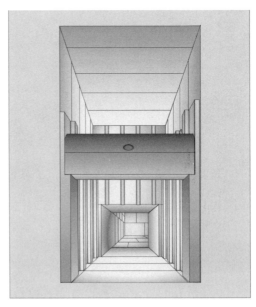

FIGURE 8. *Antechamber*

seemed to be inconsistent with any logical design for a tomb. The Grand Gallery was especially disturbing to this English mathematician and astronomer, whose mind was schooled in the orderliness of nature. He questioned that the Grand Gallery was built to serve as a stairway leading to the King's Chamber, for he had trouble making his way to the top. Its size, the corbeled walls, and the fact that it was built on such a steep angle did not indicate to him that it could have been used as a chamber either. Besides, in order to get to the Grand Gallery, he had to double over and squeeze his way through the Ascending Passage in much the same way as Al Mamun and his men did before him.

At the end of the Grand Gallery, stooping to enter the passage that led to the Antechamber, Greaves was baffled by the "portcullis" entrance, and he wondered why the walls, floor, and ceiling suddenly changed from limestone to granite. He could not even begin to fathom the complex Antechamber (see Figure 8).

The passage leading from the Antechamber to the King's Chamber is actually smaller than the sarcophagus, or coffer, that sits within the chamber, so that would had to have been installed at the time the pyramid was under construction and before the ceiling beams over the King's Chamber were put in place. Regarding the King's Chamber, Greaves wondered why a single chamber, which housed a solitary, empty coffer, needed the protection of the tremendous amount of masonry that surrounded it. He questioned why a structure as huge as the Great Pyramid was necessary for a single burial.

What is more, in the King's Chamber, Greaves observed small openings in both the north and south walls. They were not given much attention at first, and were thought to be receptacles for candles or lamps. However, after Howard-Vyse's assistant, Perring, was almost decapitated when a stone shot out of one opening and barely missed his head, it became clear that the "lamp receptacles" were actually the lower ends of shafts that ran through the body of the pyramid to the outside. The stone that almost injured Perring had evidently cleared some blockage in the shaft while it was making its way to the inner chamber, for immediately afterward a rush of cool air entered the chamber. It is reported that with the clearing of the shafts to the King's Chamber, the chamber maintained a constant temperature of 68°

FIGURE 9. *Southern Shaft in the King's Chamber*

Fahrenheit, no matter what the weather or temperature was outside. This temperature is no longer constant because the tourists who go through the Great Pyramid nearly every day generate body heat and moisture. I have been left hot and sweaty each time I crawled through the Ascending Passage or climbed the Great Gallery. The problems associated with this elevation of temperature and humidity in the King's Chamber prompted the Egyptians to contract with Rudolph Gantenbrink, a German engineer, to install fans in the Northern and Southern Shafts to improve the circulation of air (see Figure 9).

What purpose do these shafts serve? Imagine the difficulty of including these shafts in the construction of the pyramid. If they were intended to supply the King's Chamber with air, a simpler method of construction could have been used, for instance, following a horizontal path along a course of masonry to the outside. This alternative method probably would have resulted in greater airflow as well. Because of these considerations, and the fact that the dead do not breathe, Egyptologists believe that the shafts were not intended to ventilate the chamber at all, but were constructed purely for symbolic or cultic reasons.

In addition, there are significant technical problems associated with constructing the pyramid with the shafts on an incline. The limestone blocks that form the shafts on the north side would have needed precise compound angles on their adjoining faces as they turn to avoid the Grand Gallery (see Figure 10).

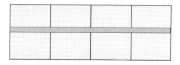

A horizontal shaft would be the simplest and most efficient way to ventilate.

When I was inside the Great Pyramid in 1995, I noticed an iron gate covering an opening in the passageway between the Grand Gallery and the Antechamber. The iron gate was unlocked, so I took the opportunity to climb into a small tunnel with my flashlight to see where it went. When I reached the end, I found myself looking at what remains of the Northern Shaft, and I was able to witness the quality of fit between those limestone blocks. As I swept the shaft with my light, I could see the fan

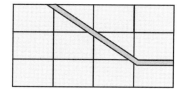

An angled shaft demands a higher degree of technical knowledge and skill.

FIGURE 10. *Horizontal Airshaft vs. Angled Airshaft*

that Gantenbrink had installed to ventilate the chamber. The shafts were exposed on the bottom side, and I was able to see that they were cleanly cut, with square, sharp inside corners. Knowing of the precision built into the rest of the pyramid, and into other ancient artifacts I had seen in Egypt, I was not surprised by the quality of workmanship—though having worked with compound angles where features of a component have to fit together without any mismatch, I could not help but be impressed. This kind of precision is not coincidental, and the builders would not have invested the resources necessary to cut and construct this feature if there was not a real need for such precision. That likelihood in itself contradicts the symbolic or cultic reasons Egyptologists ascribe to the shafts. And besides, there are simpler ways of illustrating symbolism and cultism, such as the reliefs and paintings that the ancient Egyptians created with great skill.

Any theory about the Great Pyramid should both satisfy the demands of logic and provide answers for all the relevant discoveries that have promoted so much perplexity in the past. As we have seen in this chapter, current theories regarding the function and construction of the pyramid fall ort. A credible theory would have to explain the following conditions found inside the Great Pyramid:

- The selection of granite as the building material for the King's Chamber. It is evident that in choosing granite, the builders took upon themselves an extremely difficult task.
- The presence of four superfluous chambers above the King's Chamber.
- The characteristics of the giant granite monoliths that were used to separate these so-called "construction chambers."
- The presence of exuviae, or the cast-off shells of insects, that coated the chamber above the King's Chamber, turning those who entered black.
- The violent disturbance in the King's Chamber that expanded its walls and cracked the beams in its ceiling but left the rest of the Great Pyramid seemingly undisturbed.
- The fact that the guardians were able to detect the disturbance inside the King's Chamber, when there was little or no exterior evidence of it.
- The reason the guardians thought it necessary to smear the cracks in the ceiling of the King's Chamber with cement.
- The fact that two shafts connect the King's Chamber to the outside.
- The design logic for these two shafts—their function, dimensions, features, and so forth.

Any theory offered for serious consideration concerning the Great Pyramid also would have to provide logical reasons for all the anomalies we have already discussed and several we soon will examine, including:

- The Antechamber.
- The Grand Gallery, with its corbeled walls and steep incline.
- The Ascending Passage, with its enigmatic granite barriers.
- The Well Shaft down to the Subterranean Pit.
- The salt encrustations on the walls of the Queen's Chamber.
- The rough, unfinished floor inside the Queen's Chamber.

- The corbeled niche cut into the east wall of the Queen's Chamber.
- The shafts that originally were not fully connected to the Queen's Chamber.
- The copper fittings discovered by Rudolph Gantenbrink in 1993.
- The green stone ball, grapnel hook, and cedar-like wood found in the Queen's Chamber shafts.
- The plaster of paris that oozed out of the joints inside the shafts.
- The repugnant odor that assailed early explorers.

As I have said, there are reasons for everything, and each of the above items is assuredly the effect of some cause. When searching for a solution to the enigmas of the Great Pyramid, assuming that all other explanations do not satisfy us, we must take all the evidence into consideration, even the most seemingly trivial details. In the chapters that follow, I will examine these details to prove that even information that seems unimportant may have had a most significant cause, and what has previously received just a passing look by researchers may hold the key to solving the whole problem. Because current theories do not provide satisfying answers to the questions raised by the Great Pyramid, researchers continue to cut tunnels, dig passages, and probe the pyramid, using advanced electronic sounding devices in an attempt to acquire just one more secret. As my theory will show, however, answers to much of the mystery may lie in what already has been found.

Chapter Three

PRECISION UNPARALLELED

 *fter reading considerable material on the subject of the Great Pyra-* mid and studying the drawings that accompanied the texts, it appeared to me that the opponents of the tomb theory had a valid point. With this in mind, I looked more closely at what I considered to be the most significant information regarding the Great Pyramid, which was the accuracy with which it was built. It soon became obvious to me that the researchers on both sides of the issue were sympathetic to the craftspeople involved in building the pyramids. But the researchers were not craftspeople themselves, and they did not have the perspective gained through years of experience working with their hands and with machinery. Having that experience myself, I have some very strong opinions regarding the level of manufacturing expertise practiced by the ancient Egyptians. They were not primitive by any means, and their craftsmanship and precision would be an extreme challenge to duplicate today.

During my research on the Great Pyramid, and in considering the many questions raised by others, I began to form an opinion regarding the true purpose of this structure. The decision to write this book came about after careful consideration of what courses of action were available to me to share ideas I had developed regarding the pyramid and other artifacts described by Egyptologists, especially William Flinders Petrie. As a craftsman and engineer who has worked with close tolerances for more than thirty-five years, it was only natural for me to find great affinity with the people whose remarkable accuracy was evident in building this structure.

For readers not familiar with the issues of manufacturing, let me pause briefly to provide a short historical overview. The industrial revolution, which had its genesis in England in the early 1800s, brought about standardization in the manufacture of components. Take, for instance, the rifle. At one time, each part of a rifle was manufactured and individually tailored to fit another part. There was no standardization of precision whereby interchangeable pieces could be taken off the shelf and appropriately fitted into the rifle without some adjustment. Each component was customized to fit with the other. Eli Whitney first proposed standardizing rifle components in order to facilitate supplies for war; however, in order to achieve standardization, unwelcome variations had to be worked out of the manufacturing process. In other words, it would be very unlikely that a shaft produced on a lathe that machined variations of .010 inch in diameter would precision-fit a bore with the same variations. Machines with greater precision were needed, along with a system of measurement that was standardized and closely controlled to monitor the products produced by these machines.

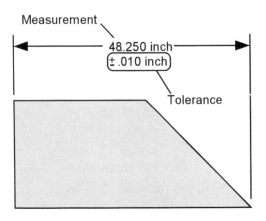

FIGURE 11. *Measurement and Tolerance*

Metrology is the science of the use of measuring equipment that is closely calibrated and monitored. The equipment requires a greater degree of precision than the object that is being produced. That being the case, we are assured that the object conforms to specification. Normally a measuring instrument, or gauge, for checking the precision of a product has a tolerance of ten percent of the tolerance of the object (see Figure 11).[1] Although the accuracy exhibited in the Great Pyramid was recorded over a century ago, it would be helpful to reevaluate the findings of early explorers in the light of today's technology.

When Petrie made his critical measurements of the Great Pyramid casing stones in 1882, he was astounded by what he found: "The eastern joint of the northern casing stones is on the top .020, .002, .045 wide; and on the face .012, .022, .013, and .040 wide. The next joint is on the face .001 and .014 wide. Hence the mean thickness of the joints is .020; and, therefore, the mean variation of the cutting of the stone from a straight line and from a true square, is but .010 on length of 75 inches up the face, an amount of accuracy equal to most modern opticians' straight edges of such length."[2]

Petrie's close examination of the casing stones revealed variations so minute that they were barely discernible to the naked eye. The records show that the

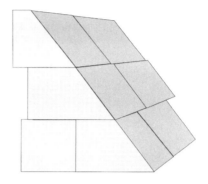

To manufacture just two blocks with a tolerance of .010 inch and place them together with a gap of no more than .020 inch is a remarkable feat. To manufacture and position over 100,000 similar blocks requires an industry that the ancient Egyptians are not credited with having developed.

FIGURE 12. *Casing Stones of the Great Pyramid*

outer casing blocks were square and flat, with a mean variation of 1/100 inch (.010) over an area of thirty-five square feet. Fitted together, the blocks maintained a gap of 0 to 1/50 inch (.020), which might be compared with the thickness of a fingernail. Inside this gap was cement that bonded the limestone so firmly that the strength of the joint was greater than the limestone itself. The composition of this cement has been a mystery for years.

The casing blocks were reported to weigh between sixteen and twenty tons each, with the largest blocks measuring five feet high, twelve feet long, and eight feet deep (see Figure 12).

It was these figures that greatly influenced my preliminary assessment of the pyramid. Here was a prehistoric monument that was constructed with such precision that you could not find a comparable modern building. More remarkable to me was that the builders evidently found it *necessary* to maintain a standard of precision that can be found today in machine shops, but certainly not on building sites.

These details are important, and we should consider them as we seek to determine how the ancient Egyptians quarried, dressed, and assembled those blocks. The general population of a century ago would not have fully appreciated the significance of such fine tolerances. At that time opticians were the only artisans who worked with such fine tolerances. Today, any researcher wishing to compare the skills found in the Great Pyramid with modern-day craftspeople would have a variety of skilled trades from which to choose.

Although the exact precision demonstrated in the manufacture and assembly of the Great Pyramid may have had little significance a century ago, there are, at this time, many people who are intimately familiar with these dimensional tolerances. I am one of them, for many years creating products with tolerances much finer than .010 inch. I know what it takes to hold such fine tolerances—and there is a great difference between knowing what .010 inch is from an abstract academic viewpoint and understanding what .010 inch is from hands-on, practical experience.

This is why I laugh when I hear intelligent men and women proposing that the pyramids and other artifacts were created using hammers and chisels. Other machinists, toolmakers, and engineers with whom I have discussed this issue are equally amused, and normally just shake their heads and mutter something straightforward and unprintable. These workers, the members of what we consider a highly advanced civilization, understand the following: It is very well to dream, speculate, and theorize, but when it comes to doing the work, we are generally brought down to earth and hard facts. The most efficient and economically minded designers and engineers are those who have experienced the manufacturing phase of their ideas and have worked on the bench and with the machines. These experiences lead them to be more realistic in their demands of skilled craftspeople.

Through my own experience in manufacturing, I have realized that theories and ideas that seemed to work fine in my mind, or on paper, could be rendered unworkable when I actually tried to apply them. In much the same way, I have found that many theories regarding the building of the Great Pyramid are not supported with material proof, for no one, despite numerous attempts, has been able to duplicate the structure using the methods theorized to have been in place in ancient Egypt. These methods have been applied, with limited success, in building smaller structures, but they are not attempts to replicate the more difficult aspects of the building. A pyramid that is twenty- or fifty-feet tall and built with limestone blocks that weigh no more than two tons does not explain how the ancient pyramid builders raised seventy-ton blocks of granite to a height of two hundred feet. Scaling up a project does not necessarily follow a linear path, nor does it rely solely on a fixed set of assumptions. So the researchers' fifty-foot pyramid may not necessarily provide them with all the data necessary to calculate the requirements for building the Great Pyramid.

Again, let us look to a technology common to our own generation to present an example of using the wrong assumptions when scaling up a project. Take, for instance, the early development of industrial lasers. As physicists, electrical engineers, optical engineers, and mechanical engineers accomplished the development of high-powered industrial lasers, they made an assumption that because the laser did not apply any mechanical force to the workpiece, the machine did not have to be as sturdy as those used in conventional machining operations—such as milling or lathe turning—where tremendous mechanical forces exert pressure on the tool and the machine. Working in the laboratory with machine members (or stages) no longer than twelve inches, researchers proved this assumption correct. However, when they built a machine that was three or four times larger, they found that other forces—such as inertia—came into play, and they realized that the machine-tools that carried these lasers had to be equally as robust and as strong as conventional machines. The situation in which Egyptologists would find themselves, I believe, would be quite similar if they scaled up their demonstration pyramid to the dimensions and precision of the Great Pyramid.

People who spend their entire careers building things, either on a building site or in a manufacturing tool shop, will know of several ways to do a task. An Egyptologist's attempt to build a pyramid using primitive means may be experimental archaeology, but because it is based on a technologically limited insight into the real significance of the Great Pyramid, it is not scientific; it only proves that what the researchers accomplished can be accomplished in the manner it was accomplished, nothing more. I applaud Dr. Mark Lehner's honesty in confessing that he had used steel tools and a front-end loader while building the demonstration pyramid for the WGBH/NOVA documentary *This Old Pyramid*.[3] I wonder, though, why that construction effort was cut from the film and we viewers never got to see it.

The most refreshing account of the talent possessed by the builders of the Great Pyramid can be seen in a video produced by Atlantis Rising Video. An interview with respected builder and architect James Hagan, who designed the Walt Disney shopping village in Lake Buena Vista, Florida, the concrete Sanford Stadium at the University of Georgia, and the impressive MARTA Five Points Central Station in Atlanta, reveals an architectural genius of modern times who uses every technique available for modern structures, yet is humbled by the creation of the Great Pyramid. Without pride or arrogance, his humility was combined with awe as he afforded the builders of the Great Pyramid the highest accolade one professional can bestow on another. "The Egyptians, or whoever built the pyramid," he said earnestly in his southern drawl, "they could build anything they want to!"[4] His comment becomes more significant when it is understood within the context he set forth, admitting that it would be impossible to build a Great Pyramid today using modern building methods, and, therefore, impossible by primitive methods. "The thing I am concerned about," he said, "are the elements of the construction, and how they came to be. These are the principles I am involved with in my world, and these are the principles I apply to the Great Pyramid." The precision built into the pyramid puzzled Hagan. He doesn't understand why this kind of precision would be necessary. Modern buildings do not require that kind of accuracy, so there is no reason to do it. "So why," he mused, "did they try to accomplish it is the first mystery." His hands-on, real-world experience is bolstered by innocent sincerity and respect that transcends the plethora of amateurs (compared to him) who profess to "know" how the pyramids of Egypt were built, and is credible support for those who still see a mystery in this edifice, and who are still seeking answers.

It is interesting how Hagan and I came to this point. In school we were both taught that the pyramids were the tombs of the Egyptian kings. After working many years in our respective technological fields, we became more aware of the construction challenges the Great Pyramid presented, and we could compare it with our practical experience. Others are not so fortunate. They must rely only on the experience and opinions of others. General knowledge of the Great Pyramid has been greatly influenced by traditional historical teachings, which are then carried into the popular media. However, the information I am focusing on in this book is not available in most popular works, and so the general population has little comparative data to work with as they evaluate Egyptologists' theories. In proposing methods of con-

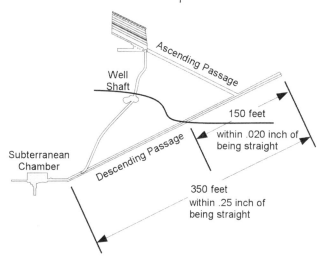

Well
Shaft

Ascending Passage

150 feet

within .020 inch of
being straight

Subterranean
Chamber

Descending Passage

350 feet
within .25 inch of
being straight

FIGURE 13. *The Descending Passage*

struction, academics have given little or no consideration to the fine tolerances maintained throughout the Great Pyramid's structure. They pass over
the astounding accuracy of the Descending Passage's construction, or at best
give it just cursory consideration (see Figure 13). These facts have not attracted the critical attention they deserve because there is a big difference
between reading these figures in a book and the actual experience of having
to maintain this precision in one's work.

Regarding the measurements taken by early explorers at the Great Pyramid and the possibility of duplicating this structure while maintaining similar
tolerances throughout, the many craftspeople with whom I have discussed
these details disavow the primitive construction methods that Egyptologists
propose. In my research, I had the opportunity to question modern stonecutters and find out the tolerances they work with. For instance, Indiana is
famous for its limestone quarries—there are approximately thirty-three of
them in and around Bedford—and they have a long history of providing
limestone for many famous buildings, most notably New York's Empire State
Building and the Waldorf-Astoria hotel.

At one time I lived sixty miles away from Bedford. One day I took an
easy and pleasant drive through the picturesque southern Indiana countryside, which was ablaze with fall foliage, to talk to Tom Adams, who at that

time worked at one of the quarries. Adams worked in the shop, cutting and dressing the stone, and the accuracy he was required to maintain in his work was not as stringent as for those who work with machine tools. Any craftspersons in a tool shop or machine shop can tell you exactly the tolerances they are working to. I asked Adams about the tolerances they work to in the quarries. He answered, "Pretty close." I asked, "How close is pretty close?" He responded, "Oh, about a quarter of an inch." Adams was astounded to hear that the limestone in the Great Pyramid was cut to .010-inch tolerance. His response regarding the abilities of the pyramid builders confirmed my belief that, contrary to what we have been taught, the pyramid builders were not primitive workers of stone.

It was clear to me that modern quarrymen and the ancient pyramid builders were not using the same set of guidelines or standards. They were both cutting and dressing stone for the erection of a building, but the ancient Egyptians somehow found it necessary to maintain tolerances that were a mere four percent of modern requirements. Two questions sprang from this revelation. Why did the ancient pyramid builders find it necessary to hold such close tolerances? And how were they able to consistently achieve them?

It goes without saying that if we were to build a Great Pyramid today, we would need a lot of patience. In preparation for his book *5/5/2000 Ice: The Ultimate Disaster,* Richard Noone asked Merle Booker, technical director of the Indiana Limestone Institute of America, to prepare a time study of what it would take to quarry, fabricate, and ship enough limestone to duplicate the Great Pyramid. Using the most modern quarrying equipment available for cutting, lifting, and transporting the stone, Booker estimated that the present-day Indiana limestone industry would need to triple its output, and it would take the entire industry, which as I have said includes thirty-three quarries, *twenty-seven years* to fill the order for 131,467,940 cubic feet of stone.[5] These estimates were based on the assumption that production would proceed without problems. Then we would be faced with the task of putting the limestone blocks in place.

The level of accuracy in the base of the Great Pyramid is astounding, and is not demanded, or even expected, by building codes today. Civil engineer Roland Dove, of Roland P. Dove & Associates, explained that .02 inch per foot variance was acceptable in modern building foundations. When I informed him of the minute variation in the foundation of the Great Pyramid, he expressed disbelief and agreed with me that in this particular phase of construction, the builders of the pyramid exhibited a state of the art that would be considered advanced by modern standards.

In *Pyramid Odyssey*, William Fix stated that the most accurate survey of the base of the Great Pyramid showed it as 3023.13 feet around the perimeter, with the average of the sides being 755.78 feet. If the alignment of this structure was governed by today's building standards, then one side of the Great Pyramid would be allowed a variation of 15.115 inches.

The generally accepted academic theory on how the base of the Great Pyramid was leveled for the most part cannot account for this accuracy. Egyptologists propose that the area was leveled through the use of standing water: A grid-like system of canals was dug into the bedrock where the Great Pyramid was to stand, and then these canals were flooded. The dry rock, or bank of the canals, was cut level, using the surface of the water as a height gauge. Although there is no evidence to support this traditional theory, at first glance it does appear to have some elements of logic. If we believe that the pyramid builders were not sufficiently advanced to have developed the precision tools that are used by today's surveyors, that would seem to be the only method available to a primitive society. However, proponents of this theory sometimes fail to mention that there is an outcrop of bedrock that was left intact at the center of the pyramid. This would mean that any grid canals would have encircled the bedrock mound.

More importantly, another detail that so far has not been given any consideration by proponents of this theory is: At what rate would the water in the grid canals system have been absorbed into the porous nummulitic limestone bedrock of the plateau, or have evaporated into the atmosphere? The grid system theory of leveling the base of the Great Pyramid is accepted on the premise that the standing water remained at a constant level in the canals. If such canals were indeed cut, how much water would be needed to reach a saturation point of the limestone plateau, which would be necessary for the water to remain at a certain level in the canals? The presence of fissures in the limestone possibly could be overcome by packing mud into them. However, this still does not explain why a primitive society, which had supposedly not yet invented the wheel, felt the need to build to such tolerances. If they did have that need and the grid system was the only method available to them, it would seem that this process would be so arduous and fraught with uncertainty that the very idea would be open to debate and promptly dismissed by the planners.

There is no evidence to support the theory that water channels facilitated the leveling of the Great Pyramid, and such a method does not seem very reliable. Mark Lehner proposed that a series of holes in the pavement around the Great Pyramid may have held sticks that were used as measuring devices, but this technique does not account for the pyramid's incredible precision. There is a modern instrument similar to that proposed by Lehner—it is called a transit, which is three sticks (a tripod) with a sophisticated measuring device on the top—but even with this instrument modern builders are not required to achieve such precision.

There is evidence that shows that ancient Egyptian builders used mechanical means to remove material in order to level the limestone bedrock for the foundation of various structures. In *Pyramids and Temples of Gizeh*, Petrie noted, "At El Bersheh (lat. 27°42') there is a still larger example, where a platform of limestone rock has been dressed down, by cutting it away with tube drills about 18 inches in diameter; the circular grooves occasionally intersecting, prove that it was done merely to remove the rock."[6]

Petrie's observations strongly support the speculation that the ancient Egyptians did not carry out their work with painstaking, back-breaking labor, but completed it with speed and precision through the employment of tools that would not be out of place on today's building sites. It certainly makes sense to cut away excess material by using a rotating "drill" and working it down to a preselected required depth. These methods of removing excess material are common in machine shops today. Therefore, it could be suggested that as far as foundation laying goes, the ancient Egyptians had reached a finite state of the art, where there was little room for improvement.

From their precisely leveled plateau, the ancient pyramid builders raised a mountain of limestone and granite with the same care and precision with which they laid its foundation. The estimated height of the Great Pyramid is 480.95 feet. It is estimated to weigh 5,300,000 tons and contain 2,300,000 blocks of stone. The stones that makes up the bulk of the pyramid are limestone, which was quarried locally on the plateau itself and in the Mokattam Hills across the Nile River, twenty miles away. The inner stones are poorer quality and are known as nummulitic (nummulitic is used to describe round fossil shells; it literally means "coin-shaped"). The composition of the stone is calcium carbonate ($CaCo_3$), which is an important fact to remember when we later look at the evidence that supports my theory.

The quantity of stone that had to be quarried, hauled, and hoisted into place in the Great Pyramid becomes even more impressive when it is compared with other civil engineering feats, whether real or imagined. It has been stated that it contains more stone than that used in all the churches, cathedrals, and chapels built in England since the time of Christ. Thirty Empire State Buildings could be built with the estimated 2,300,000 stones. A wall three-feet high and one-foot thick could be built across the United States and back using the amount of masonry contained in the Great Pyramid. The list of such comparative observations is long and could fill many pages, but these few suffice to impress upon us the prodigious feat the ancient builders accomplished.

The Great Pyramid's orientation is as impressively precise as its construction. It is oriented within three minutes of a degree from true north. Researchers speculate that because the pyramid was built 4,800 or more years ago, this variation may have been caused by a shifting of the Earth's crust or of its axis. Whatever the reason for its slight deviation from absolute true north, the Great Pyramid was the most accurately aligned structure in the world until the building of the Paris Observatory.

Adding to the mystery of the Great Pyramid is the fact that its shape appears to incorporate the mathematical function of pi. This incommensurable number, 3.14159 ad infinitum, exists in a pyramid when the angle of the pyramid's sides is 51°51'14" per side. Given such an angle, the perimeter of the pyramid is in relationship to its height as the circumference of a circle is to its radius. It may be stretching the truth a little to say that the Great Pyramid had this exact angle, or that it was the builders' expressed intention to have a structure that exhibited this mathematical constant. Still it was certainly close. Petrie's measurements unequivocally show that the angle of the sides of the pyramid was constructed with remarkable precision. He wrote, "On the whole, we probably cannot do better than take 51°52' plus or minus 2' as the nearest approximation to the mean angle of the pyramid, allowing some weight to the South side."[7] Having worked with blueprints where tolerances on angles are frequently given as plus or minus one degree unless otherwise specified, I am certain that Petrie's measurements indicate that the angle of the Great Pyramid was a critical part of the entire structure.

As we can see, there is four minutes of a degree tolerance band within which anybody so desiring could arrive at the perfect pi angle of 51°51'14". This angle fits well within the tolerance band described by Petrie, and if we wanted to choose this particular figure to prove that the builders had the knowledge of pi, we could probably do so. I prefer to present the data as Petrie did, with deviations that are bound to arise over such a large area. Although the incorporation of pi into the shape of the Great Pyramid has been attributed by some to be pure chance, the fact that such an angle was discovered in the casing stones suggests that the builders were at least knowledgeable in the sciences of mathematics, trigonometry, and geometry.

The enigmatic Great Pyramid initiates many very basic questions. Why is it so big? Why was it necessary to build it with such a high degree of accuracy? How was it built? Methods of transporting materials to the building site are still under debate. There have been attempts to vindicate traditional theories by following the methods that were proposed in building the pyramids. However, it could be said that hauling or dragging blocks of stone over the desert floor—just to prove that it can be done—does as much to prove that this was the way the blocks were moved as the apprentice toolmaker perspiring over his deburring work proves that his efforts explain the entire General Motors operations, or the Nippon Corporation of Japan.

It is difficult to ascertain what the Japanese Nippon Corporation was trying to prove when, in 1978, they attempted to erect a sixty-foot pyramid in Egypt. Under prescribed conditions, they received permission from the Egyptian government to erect a pyramid southeast of Mycerinus' Pyramid on the Giza Plateau. They were to use the same methods that the original pyramid builders supposedly used. They were not to use the stone from the plateau itself, but from the quarry that provided the original blocks. The rules were that after Nippon had finished this demonstration, they were to dismantle their pyramid and restore the site to its original condition.

Agreeing to these stipulations, the Japanese set to work quarrying, hauling, and erecting approximately one-ton blocks of limestone. The reports and films taken of this operation reveal the difficulties they encountered during their long and difficult task. Reportedly, their first difficulty was getting the stones across the Nile River. In *Pyramid Prophecies,* Max Toth wrote:

> *Once cut into approximate one-ton blocks, the stones could not be barged across the River Nile. Floatation apparently was not the simple answer, as had been suggested. The blocks finally had to be ferried across by steamboat.*
>
> *Then, teams of one hundred workers each tried to move these stones over the sand—and they could not move them even an inch! Modern construction equipment had to be resorted to, and once again, when the blocks of stone were finally brought to the building site, the teams could not lift their individual stones more than a foot or so. In the final construction step, a crane and helicopter were used to position the blocks.*[8]

The reported difficulty the Japanese encountered moving the limestone seems to conflict with other reports by Egyptologists. According to I.E.S. Edwards, moving a one-ton block of limestone was not as difficult as the Japanese made it out to be:

> *Without wheeled vehicles, how did the Egyptians move such great weights, and how did they raise them to the heights of the pyramids at Giza? In an experiment not many years ago, a French investigator obtained a one-ton block of limestone. The block was positioned on a track of moist mud taken from the Nile, and a crew of about 50 men was assembled and instructed to pull it with ropes. When they started pulling, the block slid along as if it were almost weightless. With half as many men, the block again ran away. The experimenter soon found that one man, with no difficulty, could push the ton of stone along on the wet mud. In ancient records we see Egyptian crews dragging great weights on sledges while waterbearers wet down the surface ahead.*[9]

Being objective about the whole debate of whether the methods of building the pyramids were those proposed by Egyptologists or other more advanced methods, and considering the experiments undertaken to obtain conclusive proof one way or another, we would have to question the motives of the experimenters. It seems that if the Japanese team were out to prove that pyramid building should be left to those with access to modern techniques and equipment, they succeeded. But this does not mean that their conclusions were irrefutable, and it does appear that they gave up easily when trying to move *their* one-ton blocks of limestone. However, the records show that they did attempt the construction of a pyramid (one that was small compared to the originals) and failed to complete it using primitive methods. It now remains for those who are absolutely convinced that the ancient Egyptians constructed the pyramids using primitive techniques to build a pyramid themselves, using those same techniques that they propose the Egyptians used. As part of such an attempt, it would help if they cut out just one seventy-ton block of granite from the Aswan quarry, which is located five hundred miles away, using their hardened copper chisels or dolerite balls and then transported the block to the Giza Plateau with their barges, ropes, and manpower. If the proponents of traditional theories of constructing the pyramids are able to accomplish this feat, then we should give serious consideration to their proposals about pyramid construction.

A more recent attempt at building a pyramid was carried out by a team from the television program *NOVA* that included Egyptologist Mark Lehner and Massachusetts stone mason Roger Hopkins. Their pyramid reached a height of twenty feet and took three weeks to build using steel tools, and front-end loaders, and the valiant effort of laborers, who, under the direction of Hopkins and for the benefit of the cameras dragged the final stones into place. As Dr. Lehner honestly described the small scale of this undertaking, "It would have fitted neatly on to the top of the Great Pyramid, in whose shadow we built it."[10]

Considering the immense size of the Great Pyramid, the precision with which it was built, the materials that were used, and the uniqueness of its interior passages and chambers, we are faced with a structure that has no parallel in modern times. Theories of primitive methods of construction are invalidated by the proponents' own tests and demonstrations; and considering the time in which the proponents of these theories have had to develop a watertight case, it appears they will never figure it out. The passing years have significantly weakened the traditional views, perhaps not in the eyes of Egyptologists, but certainly in the eyes of people like myself. We are beginning to learn the true significance of alternative theories that have, for the most part, been ignored. Meanwhile, credibility of the old theories has been undermined by highly questionable armchair speculations that are passed on as facts—such as the idea that the Egyptians used copper chisels to shape hard, igneous rock.

The most compelling evidence for the likelihood that the Great Pyramid was constructed by craftspeople with specialized knowledge and advanced techniques is the precision with which it was built. This precision reveals more about the true nature of its builders than any inscription or cartouche. There is no way to ignore the accuracy of this stonecutting, despite Egyptologists' interpretations of the inscriptions found in pyramids or temples in Egypt. After all, hieroglyphics, like any language, has the potential to be misunderstood.

After discussing much of the preceding information with the artisans at today's building sites, machine shops, and quarry mills, I became aware of the reason why we are still influenced by ideas that are not compatible with practical application. The artisans of today are too busy making a living to give serious thought to scholarly theories, and even when gross inequities are presented to them, they respond with a cynical shrug. When told that giant limestone casing stones, which were cut to within 1/100 of an inch, were cut with hammer and chisel, a typical response was a shake of the head.

As for the general public's lack of interest in the technological mysteries surrounding the Great Pyramid, the fact is that the majority enjoy the use of technology without regard to its creation. We buy tools, utensils, and appliances with little thought about the skill and ingenuity that played a part in the production of just one little component. We appreciate the final product but have little knowledge of how it was made.

Because I have had technical experience in the procedures that created some modern conveniences, I have examined the evidence in a new way and I sum up my thoughts regarding the construction of the Great Pyramid as follows: It has been said many times that the Great Pyramid was built with tolerances that modern opticians would be hard-pressed to match. In ana-

reason for this high degree of perfection, I consider two possible answers. First, the building was for some reason *required* to conform to precise specifications regarding its dimensions, geometric proportions, and its mass. As with a modern optician's product, any variation from these specifications would severely diminish its primary function. In order to comply with these specifications, therefore, greater care than usual was taken in manufacturing and constructing this edifice. Second, the builders of the Great Pyramid were highly evolved in their building skills and possessed greatly advanced instruments and tools. The accuracy of the pyramid was normal to them, and perhaps their tools were not capable of producing anything less than this superb accuracy, which has astounded many over the years. Consider, for example, that the modern machines that produce many of the components that support our civilization are so finely engineered that the most inferior piece they could turn out is more accurate than what was the norm for those produced one hundred years ago. In engineering, the state of the art inevitably moves forward.

Of course, there might be a third alternative. It could always be said that those who built the Great Pyramid did not really know what they were doing, and that the end result of their labors was achieved purely through trial and error—the Great Pyramid's precision was just a stroke of luck. Yes, I know that reasoning sounds ridiculous, but it has been suggested by Egyptologists and other researchers, actually more than a few times. If we pass the achievement of the builders of the Great Pyramid off as pure coincidence, we need say little more. However, if curiosity gets the better of us, we can look a little more closely and consider the notion that perhaps there is some significance behind the Great Pyramid's mathematical sophistication.

Today we do not invest the time and effort to finish modern artifacts to a precision of .0002 inch, unless we must. As a general rule, when estimating the cost of manufacturing an object, if the tolerance box has an extra zero in it—that is, not .001 but .0001—the price of the object increases significantly. The more precise the object, the more it costs, because the labor is more expensive. Tool and instrument makers earn more money per hour than machine operators. Therefore, it follows that for cost purposes, an engineer will design a machine or tool with the greatest amount of tolerance allowable while at the same time still providing for the machine's or tool's proper function. The measurements of any given object are simply a means to an end, and, while being related to the object, they are not the object itself. It is not unreasonable for us to assume, then, that the dimensions and precision embodied in the Great Pyramid are means by which the builders created a product that had to function in a way that a product of lesser precision could not.

If the builders were intelligent enough to figure out the engineering aspects of quarrying, hauling, and erecting millions of tons of masonry with such precision, then is it logical to assume that this feat was achieved with primitive methods? Would the degree of practicality evident in the structure have been limited to just the comprehension of its final form, or would it have been applied to solving all constructional aspects? Wouldn't any group of individuals who could build such an advanced and unique product have the capacity to develop advanced and unique tools to produce it? With the Great Pyramid, the achievement of its final form was undoubtedly a group effort; and since the evidence available suggests the builders used sophisticated methods of machining, they would have developed the necessary advanced and unique tools to do the job. With this in mind, perhaps we should speculate that they also had an advanced and unique purpose for building the Great Pyramid.

As I have suggested, if we look at the history of manufacturing, we will find that the evolution of machine tools has resulted in a quality of product that was not possible one hundred years ago. The accuracy and repeatability of these machine tools is such that the accuracy and replicability of some of the work they produce may not be necessary for the product to function properly. The machines are built to produce highly accurate and consistent products without regard to the level of importance each feature may have in the final product. One viewpoint may be, therefore, that the pyramid builders had created machinery with a "state of the art" in cutting and dressing stone that was incapable of producing low-quality work. This may seem a far-fetched idea on the surface, but as I will show in the next chapter, advanced methods of machining stone are clearly evident in artifacts from that period.

With the Great Pyramid, we are faced with an artifact that exhibits a state of the art in manufacturing and construction that we do not find necessary for specification of modern buildings. In fact, artisans who provide materials and erect modern buildings do not even relate to the tolerances that must have been imposed on the creators of the Great Pyramid. It was with this realization that I continued my study and tried to imagine what re-creating it would take. The Great Pyramid speaks of a highly skilled and intelligent body of people who conceived and executed a design with an attention to detail that is utterly astounding. A tremendous amount of re-sources must have been made available for it. Graham Hancock said it very nicely in a documentary I had taken part in. "The builders of the pyramids speak to us across the centuries and say 'We are not fools. . . . Take us seriously!'"[11] His comment sums up exactly the conclusions I had reached in 1977: The pyramid builders were as intelligent as we are. How they applied their knowledge may have been different, but it is obvious that they possessed sufficient knowledge to create an artifact having a distinct feature that, so far, we have not been able to repeat. The bald fact is that the Great Pyramid—by any standard old or new—*is the largest and most accurately constructed building in the world.*

The discoveries at the Great Pyramid that have most interested me are those that involve the methods the ancient builders used to cut the material used to construct it—primarily the granite. Because I am involved in manu-facturing, I have noticed many inconsistencies between what Egyptologists have taught regarding the tools that were supposedly used and the evidence that can be drawn from the masonry itself. In other words, the stones of the Great Pyramid tell me a different story than they have other observers. The stones tell me that they were cut using *machine-power*, not manpower as orthodox Egyptologists theorize.

Chapter Four

ADVANCED MACHINING
IN ANCIENT EGYPT

 *n August 1984,* Analog *magazine published my article, "Advanced* Machining in Ancient Egypt?" It was a study of *Pyramids and Temples of Gizeh* by Sir William Flinders Petrie. Since the article's publication, I have visited Egypt twice, and with each visit I leave with more respect for the ancient pyramid builders. While in Egypt in 1986, I visited the Cairo Museum and gave a copy of my article, along with a business card, to the director of the museum. He thanked me kindly, threw it in a drawer to join other sundry material, and turned away. Another Egyptologist led me to the "tool room" to educate me in the methods of the ancient masons by showing me a few cases that housed primitive copper tools. I asked my host about the cutting of granite, for that was the focus of my article. He explained that the ancient Egyptians cut a slot in the granite, inserted wooden wedges, and then soaked them with water. The wood swelled, creating pressure that split the rock. Splitting rock is vastly different than machining it, and he did not explain how copper implements were able to cut granite, but he was so enthusiastic with his dissertation that I did not interrupt. To prove his argument, he walked me over to a nearby travel agent, encouraging me to buy airplane tickets to Aswan, where, he said, the evidence is clear. I must, he insisted, see the quarry marks there, as well as the unfinished obelisk (see Figure 14).

Dutifully, I bought the tickets and arrived at Aswan the next day. (After learning some of the Egyptian customs, I got the impression that this was not the first time that my Egyptologist friend had made that trip to the travel agent.) The quarry marks I saw there did not satisfy me that the methods conventional theorists describe were the only means by which the pyramid builders quarried their rock (see Figure 15). For example, located in the chan-

FIGURE 14. *Quarry Marks at Aswan*

nel which runs the length of the estimated 3,000-ton obelisk, I saw a large, conical hole drilled into the bedrock hillside that measures approximately twelve inches in diameter and three feet deep. The hole was drilled at an angle, with the top intruding into the channel space (see Figure 16). To my eye, it seemed likely that the ancients might have used drills to remove material from the perimeter of the obelisk, knocked out the webs between the holes, and then removed the cusps.

The Aswan quarries *were* educational, although after returning to Cairo the following day and while strolling around the Giza Plateau later in the week, I started to question the quarry marks at Aswan even more. South of the Second Pyramid I found an abundance of quarry marks of similar nature. The granite casing stones that had sheathed the Second Pyramid were stripped off and lying around the base in various stages of destruction. Some of the stones were still in place, though sections had been split away from them, and there I found the same quarry marks that I had seen earlier in the week at Aswan. This was puzzling to me. Disregarding the impossibility of Egyptologists' theories on the ancient pyramid builders' quarrying methods, I wondered if these theories were valid even from a nontechnical, logical viewpoint. If those quarry marks distinctively identify the people who created the pyramids, why would they engage in such a tremendous amount

FIGURE 15. *Quarry Marks on Khafre's Pyramid Granite*

FIGURE 16. *Drill Hole at Aswan*

of extremely difficult work only to destroy the work after having completed it? It seemed to me that the quarry marks at Aswan and on the Giza Plateau were made at a later time and that they were created by people who were interested only in obtaining granite without caring about its source.

As I pondered this revelation, I wondered about William Flinders Petrie, who had walked this plateau one hundred years before me. What drove him? What were his private thoughts about his studies that he did not share with the Royal Society or his colleagues? Being a pioneer in the field of Egyptology, his work greatly influenced the archaeological profession. It goes without saying that archaeology is largely the study of history's toolmakers, and that archaeologists understand a society's level of advancement from its tools and artifacts. The hammer was probably the first tool ever invented, and hammers have forged some elegant and beautiful artifacts. Ever since humans first learned that they could effect profound changes in their environment by applying force with a reasonable degree of accuracy, the development of tools has been a continuous and fascinating aspect of human endeavor. The Great Pyramid, however, leads a long list of artifacts that have been misunderstood and misinterpreted by archaeologists, who have promoted theories and methods about its construction that cannot be explained using the tools they have excavated.

For the most part, archaeologists consider the primitive tools they discover as contemporaneous with the artifacts of the same period. During the pyramid building period in Egyptian history, artifacts were produced in prolific numbers and a great many have survived—but there are precious few tools that survive to explain their creation. Consequently, the ancient Egyptian artifacts cannot be explained in simple terms. What is more, the tools that *have* survived do not fully represent the state of the art that is evident in the artifacts themselves. The tools displayed by Egyptologists as instruments for the creation of many of these incredible artifacts are physically incapable of reproducing them. After standing in awe before these engineering marvels, and then being shown a paltry collection of copper implements in the tool case at the Cairo Museum, I came away bemused and frustrated. In spite of ancient Egypt's most visible and impressive monuments, we have only a sketchy understanding of the full scope of its technology.

Petrie recognized that these tools were insufficient to explain Egyptian artifacts. He explored this anomaly thoroughly in his book, and he expressed amazement about the methods the ancient Egyptians used to cut hard, igneous rocks. He credited these ancient craftspeople with methods that we are only now beginning to appreciate. So why do modern Egyptologists insist that this work was accomplished with a few primitive copper instruments?

I am not an Egyptologist, I am a technologist. I do not have much interest in who died when, whom they may have taken with them, and where they went. I intend no disrespect for the mountain of work and the millions of hours of study conducted on this subject by intelligent scholars (professional and amateur), but my interest—thus my focus—is elsewhere. When I look at an artifact to investigate how it was manufactured, I am not concerned about its history or chronology. Having spent most of my career working with the machinery that actually creates modern artifacts—such as jet-engine components—I am able to analyze and determine how an artifact was created. I also have had training and experience in various nonconventional manufacturing methods, such as laser processing and electrical discharge machining. Having said this, I should state that, contrary to some popular speculations on the cutting of stone for the pyramids, I have *not* seen evidence of laser cutting on the Egyptian rocks. A variety of people have speculated that to erect a structure as perfect as the Great Pyramid, the builders must have possessed supernatural powers. Some even speculate that the builders used lasers to cut the masonry and then levitated the stones into place in the pyramid. While I cannot speak authoritatively regarding the builders' powers of levitation—whether the implementation of those powers was through the use of the mind or through the use of technology—I can say with reasonable confidence that no lasers were used in cutting the materials that went into building the Great Pyramid. Although the laser is a wonderful tool with many uses, its function as a cutting tool is limited to economically viable applications, such as cutting small holes in thin pieces of metal and refractory material. As a general purpose cutting tool, it cannot compete with the machining methods that were available before its inception.

Still, there *is* evidence for other nonconventional machining methods, as well as more sophisticated, conventional type sawing, lathing, and milling practices. Undoubtedly some of the artifacts that Petrie studied were produced using lathes. There is also evidence of clearly defined lathe tool marks on some "sarcophagi" lids. The Cairo Museum alone contains evidence—once it is properly analyzed—that is sufficient to prove that the ancient Egyptians used highly sophisticated manufacturing methods. For generations scholars have focused on the nature of the cutting tools used. But while in Egypt in February 1995, I, as a technologist, discovered evidence that raises a perhaps more intriguing question: "What guided those cutting tools?"

The methods used to cut the masonry for the Great Pyramid can be deduced from the marks they left behind on the stone. The bulk of the pyramid was constructed with limestone blocks weighing an average of two-and-one-half tons each. While there are some interesting points to be made concerning the limestone that encased the Great Pyramid, and they will be addressed later, those stones do not offer the same information about the methods that were used to produce them as do the thousands of tons of granite. At the expense of considerable time and effort by the original creators, the granite artifacts found in the Great Pyramid and at other sites in Egypt offer the clues we are seeking.

But before we investigate the granite that was used in the Giza pyramids, we must evaluate several artifacts that almost undeniably indicate machine power was used by the pyramid builders. Those artifacts, scrutinized by Petrie, are all fragments of extremely hard, igneous rock. Those pieces of granite and diorite exhibit marks that are the same as those that result from cutting with modern machinery. It is shocking that Petrie's studies of those fragments have not attracted greater attention, for there is unmistakable evidence of machine-tooling methods. It will probably surprise many people to know that evidence proving that the ancient Egyptians used tools such as straight saws, circular saws, and even lathes *has been recognized for over a century.* The lathe is the father of all machine tools in existence, and Petrie submitted evidence showing that the ancient Egyptians not only used lathes, but they performed tasks that would, by today's standards, be considered impossible without highly developed specialized techniques, tasks such as cutting concave and convex spherical radii without splintering the material.

While digging through the ruins of ancient civilizations, would archaeologists instantly recognize the work of machine tools by the kind of marks made on the material or the configuration of the piece at which they were looking? Fortunately, one archaeologist had the perception and knowledge to recognize such marks, although at the time Petrie's findings were published the machining industry was in its infancy. The growth in the industry since then warrants our taking a new look at his findings. (See Appendix A for an excerpt from Petrie's *Pyramids and Temples of Gizeh* regarding this topic.)

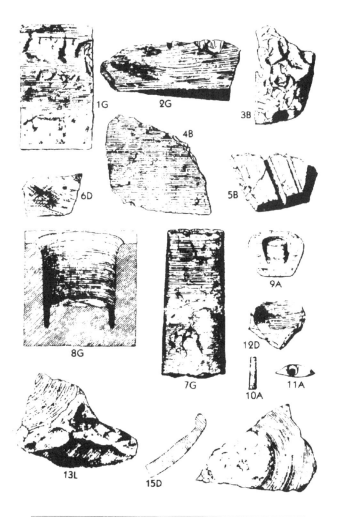

FIGURE 1. A—Alabaster B—Basalt D—Diorite G—Granite L—Limestone
—From Pyramids and Temples of Gizeh, by William Flinders Petrie, copyright 1883,
London. Reprinted by courtesy of Ann F. Petrie. A reprint edition of the book is
planned by Akadem. Druck-u. Verlagsanstalt, Graz, Austria.

FIGURE 17. *Petrie's Samples of Machining*

One can gather by reading Petrie's work that he involved himself in some extensive research regarding the tools that were employed in cutting hard stone (see Figure 17). Even so, there is a persistent belief among some Egyptologists that the granite used in the Great Pyramid was cut using copper chisels. I.E.S. Edwards, British Egyptologist and the world's foremost expert on pyramids, said, "Quarrymen of the Pyramid age would have accused Greek historian Strabo of understatement as they hacked at the stubborn granite of Aswan. Their axes and chisels were made of copper hardened by hammering."[1]

Having worked with copper on numerous occasions, and having hardened it in the manner suggested above, I was struck that this statement was entirely ridiculous. You can certainly work-harden copper by striking it repeatedly or even by bending it. However, after a specific hardness has been reached, the copper will begin to split and break apart. This is why, when copper is worked to any great extent, it has to be periodically annealed, or softened, in order to keep it in one piece. Even after being hardened in this manner, the copper is not capable of cutting granite. The hardest copper alloy in existence today is beryllium copper. There is no evidence to suggest that the ancient Egyptians possessed this alloy, but even if they did, even this alloy is not hard enough to cut granite. Yet copper has been described as the only metal available to the craftspeople building the Great Pyramid. Consequently, it would follow that all work must have sprung from their use of this basic metallic element. Theorists may be entirely wrong, however, even in their basic assumption that copper was the only metal available to the ancient Egyptians.

Another little known fact about the pyramid builders is that they were iron makers as well. You will not find much reference to this fact in textbooks; as researchers have only found one piece of wrought iron, and because of its singularity, Egyptologists have not attached much significance to it. Howard-Vyse's assistant, J. R. Hill, discovered this wrought iron within one of the joints of the Great Pyramid's limestone masonry in 1837 (see Appendix B). From there it was delivered to the British Museum. As it was the only piece of iron ever found from that era, its impact was not significant enough to change our concept of world history. However, it is important to note that if there was an abundance of iron or steel at the time of the

Great Pyramid's building, its survival would be dependent on some kind of sanctuary from the elements, such as being buried in the limestone of the pyramid. Recent analysis of this metal discovered that it had traces of gold on one surface, as though it had been gold plated at one time.

Because of the convincing documentation of the discovery of wrought iron at the Great Pyramid and the identification of the builders of the Great Pyramid as the makers of that iron, we are left to ponder the possibility that other ferrous materials existed in prehistory. It is fair of us to ask, therefore, what other kinds of metal components—without the protection of several feet of limestone—would have rusted or have been sand-blasted away during the passage of thousands of years? Without going back in time and interviewing the craftspeople who worked on the pyramids, we will never know for sure what materials their tools were made of. Any debate of the subject would be futile, for until the proof is at hand, we can reach no satisfactory conclusion. However, we can surmise the manner in which the masons used their tools, and if we compare current methods of cutting granite with the finished product (i.e., the granite coffer), we can find a solid base on which to draw several enlightening parallels.

So let us do just that. Today's granite-cutting methods include the use of wire saws and an abrasive, usually silicon-carbide, which has a hardness comparable with diamond and, therefore, is hard enough to cut through the quartz crystal in the granite. The wire is a continuous loop that is held by two wheels, one of the wheels being the driver. Between the wheels, which can vary in distance depending on the size of the machine, the granite is cut by being pushed against the wire or by being held firmly and allowing the wire to feed through it. The wire does not actually cut the granite, but is designed to effectively hold the silicon-carbide grit that in fact does the cutting. By looking at the shapes of the cuts that were made in the basalt items 3b and 5b, as shown in Figure 17, one could certainly speculate that a wire saw had been used and left its imprint in the rock. The full radius at the bottom of the cut is exactly the shape that would be left by such a saw.

Wanting to know more about the sawing of granite, I consulted John Barta, of the John Barta Company, who informed me that the wire saws used in quarry mills today cut through granite with great rapidity. In fact, Barta told me that wire saws with silicon-carbide cut through the granite like it is butter. Out of interest, I asked Barta what he thought of the copper chisel theory proposed by Egyptologists. Suffice it to say that Barta, being from Cleveland, and possessing an excellent sense of humor, came forth with some jocular remarks regarding the practicality of such an idea.

If the ancient Egyptians had indeed used wire saws for cutting hard rock, our next question would be to ask if these saws were powered by hand or machine. With my experience in machine shops and the countless number of times I have had to use saws (both handsaws and power saws), I was able to recognize strong evidence that, in at least some instances, the latter method was used.

Once again, Petrie provided us with a clue: "On the N. end [of the coffer] is a place, near the west side, where the saw was run too deep into the granite, and was backed out again by the masons; but this fresh start they made still too deep, and two inches lower they backed out a second time, having cut out more than .10 inch deeper than they had intended. . . ."[2]

The above was Petrie's notes on the coffer inside the King's Chamber in the Great Pyramid. The following concerned the coffer inside the Second Pyramid: "The coffer is well polished, not only inside but all over the outside; even though it was nearly all bedded into the floor, with the blocks plastered against it. The bottom is left rough, and shows that it was sawn and afterwards dressed down to the intended height; but in sawing it the saw was run too deep and then backed out; it was, therefore, not dressed down all over the bottom, the worst part of the sawing being cut .20 inch deeper than the dressed part. This is the only error of workmanship in the whole of it; it is polished all over the sides in and out, and is not left with the saw lines visible on it like the Great Pyramid coffer."[3]

Petrie estimated that a pressure of one to two tons on jeweled-tipped bronze saws would have been necessary to cut through the extremely hard granite. If we agree with those estimates as well as with the methods proposed by Egyptologists regarding the construction of the pyramids, then a severe inequity can be discerned between the two theories.

So far, Egyptologists have not given credence to any speculation that suggests that the builders of the pyramid might have used machines instead of manpower in this massive construction project. In fact, they do not give the pyramid builders the intelligence to have developed and used the simple wheel. It is quite remarkable that a culture that possessed sufficient technical ability to make a lathe and progressed from there to develop a technique that enabled them to machine radii in hard diorite would not have thought of the wheel before then.

Petrie logically assumed that the granite coffers found in the Giza pyramids were marked prior to being cut. The workers were given a guideline with which to work. The accuracy exhibited in the dimensions of the coffers confirms this, plus the fact that guidelines of some sort would have been necessary to alert the masons of their error.

While no one can say with certainty how the granite coffers were cut, the saw marks in the granite have certain characteristics which suggest that they were not the result of hand sawing. If there was not evidence to the contrary, I might agree that the manufacturing of the granite coffers in the Great Pyramid and the Second Pyramid could quite possibly have been achieved using pure manpower—and a tremendous amount of time. But it is extremely unlikely that a team of masons operating a nine-foot handsaw would be cutting through hard granite fast enough that they would pass their guideline before noticing the error. To then back the saw out and repeat the same error, as they did on the coffer in the King's Chamber, does nothing to confirm the speculation that this object was the result of handwork.

When I read Petrie's passage concerning these deviations, a flood of memories came to me of my own history with saws, both power and manual driven. My experience, plus my observations of others using power saws, makes it inconceivable to me that manpower drove the saw that cut the granite coffers. While cutting steel with handsaws, workers would be able to see the direction the saw blade would make well in advance of making a serious mistake, especially in an object that has a long workface and, certainly, in one with such dimensions as the coffers, which could not be cut with great rapidity. The smaller the workpiece, naturally, the faster the blade would cut through it. On the other hand, if the saw was mechanized and was cutting rapidly through the workpiece, the saw could "wander" from its intended course and cut through the guideline at a certain point at such a speed that the error could be made before the condition could be corrected. This is not uncommon. I do not mean to imply that a manually operated saw cannot wander, only that the speed of the operation would determine the efficiency in discovering any deviation that the saw might have from its intended course.

Another interesting point to consider in Petrie's observations of the coffer was that the saw was run too deeply, backed out, and then cut into the stone again. Anyone who has been faced with the problem of drawing a saw blade out of a cut and then making a restart on only one side of the cut, which is essentially what was done with the granite, knows that excessive pressure on the blade would force it back into the original cut. To make a restart of this type, it is necessary that very little pressure be put on the blade. With these considerations, it is doubtful if we can ever verify Petrie's deductions that two or three tons of pressure were necessary to cut the granite.

Making a restart in the middle of a cut, especially one of such dimensions as the granite coffer, would be more easily accomplished with machine sawing than it would be with hand sawing. With hand sawing there is little control over the blade in a situation like this, and it would be difficult to accurately gauge the amount of pressure needed. Also, the blade of the handsaw would be moving quite slowly, a fact that causes me to question further the suggestion that a handsaw was being used. I believe, based on my own and others' experiences, that the accomplishment of such a feat, using a saw at such a slow speed and with very little pressure, would be almost, if not completely, impossible.

With a power-driven saw, on the other hand, the blade moves rapidly and can be more easily controlled. The blade can be held in a fixed position, with uniform pressure over the entire length of the blade, and in the direction necessary to restart a cut. A worker can accurately maintain this front and side pressure until sufficient material has been removed from the workpiece to allow a continuation of normal cutting speed. The evidence from the Great Pyramid's granite coffer shows that workers attained a normal cutting rate shortly after they rectified the mistake, a fact that can be deduced by noting that the mistake was repeated two inches further along. At that point the blade was cutting through the granite at the wrong place faster than the workers were able to detect and stop it.

Another method of correcting a mistake while using a handsaw, if the error was only in a small area of the cut, would be to tilt the blade and continue cutting in the unspoiled area, so that when the blade reached the area that needed correcting, it would be supported by the fresh tilted cut and would have sufficient strength to combat any tendencies to follow the original straight cut. If the granite coffer had been cut with handsaws, it is conceivable that the workers could have used this method to correct their cutting errors. However, it has probably become apparent by now that Petrie had the eye of a hawk and documented just about everything he saw. At the same time he was studying the cutting mistakes in the granite, he also noticed other features of these artifacts: "[The coffer in the King's Chamber] is not finely wrought, and cannot in this respect rival the coffer in the Second Pyramid. On the outer sides the lines of sawing may be plainly seen: horizontal on the N., a small patch horizontal on the E., vertical on the S., and nearly horizontal on the W.; showing that the masons did not hesitate at cutting a slice of granite 90 inches long, and that the jeweled bronze saw must have been probably about 9 feet long."[4]

If the operators of the saw, in an attempt to correct a mistake, had tilted their blade in the manner described above, the saw lines would show a difference from the pre-error saw lines because they would be at an angle. The mistakes in the granite were found on the north side of the coffer, and Petrie observed that the saw lines on that side were horizontal. Following Petrie's footsteps, in 1986 I was able to verify his observations of the coffer in the Great Pyramid. The saw lines on the side where the mistakes were made are all horizontal, invalidating any argument proposing that the mistake was overcome by tilting the blade, which is probably the only method that would be successful using a handsaw. This evidence points to the distinct probability that the pyramid builders possessed motorized machinery when they cut the granite found inside the Great Pyramid and the Second Pyramid.

Today these saw marks would reflect either the differences in the aggregate dimensions of a wire bandsaw with the abrasive, or the side-to-side movement of the wire, or the wheels that drive the wire. The result of any of these conditions is a series of slight grooves. The feedrate and either the distance between the variation in length of the saw or the diameter of the wheels determine the distance between the grooves. The distance between the grooves on the coffer inside the King's Chamber is approximately .050 inch.

Along with the evidence on the outside of the King's Chamber coffer, we find further evidence of the use of high-speed machine tools on the

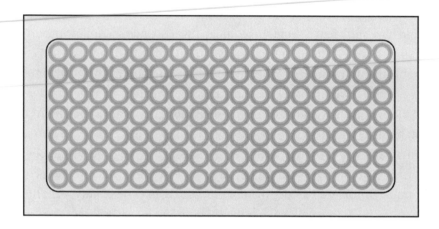

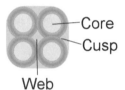 Core
Cusp
Web

After tubular drilling a box, there are cores and webs that need to be removed. Following their removal, the cusps that remain are machined away until the surface is flat.

FIGURE 18. *Tube-drilling a Granite Box*

inside of the granite coffer. The methods that were evidently used by the pyramid builders to hollow out the inside of the granite coffers are similar to the methods that would be used to machine out the inside of components today. Tool marks on the coffer's inside indicate that when the granite was hollowed out, workers made preliminary roughing cuts by drilling holes into the granite around the area that was to be removed (see Figure 18). According to Petrie, those drill holes were made with tube-drills, which left a central core that had to be knocked away after the hole had been cut. After all the holes had been drilled and all the cores removed, Petrie surmised that the coffer was then handworked to its desired dimension. The machinists on that particular piece of granite once again let their tools get the better of them, and the resulting errors are still to be found on the inside of the coffer. As Petrie noted, "On the E. inside is a portion of a tube-drill hole remaining, where they had tilted the drill over into the side by not working it vertically. They tried hard to polish away all that part, and took off about 1/10 inch thickness all around it ; but still they had to leave the side of the hole 1/10 deep, 3 long, and 1.3 wide ; the bottom of it is 8 or 9 below the original top of the coffer. They made a similar error on the N. inside, but of a much less extent. There are traces of horizontal grinding lines on the W. inside."[5]

The errors Petrie noted are not uncommon in modern machine shops, and I must confess to having made them myself on occasion. Several factors could be involved in creating this condition, although I cannot visualize any one of them being a hand operation. Once again, while working their drill into the granite, the machinists had made a mistake before they had time to correct it.

Let us speculate for a moment that the drill was being worked by hand. How far into the granite would the Egyptian craftspeople have been able to cut before the drill had to be removed to permit cleaning the waste out of the hole? Would they be able to drill eight or nine inches into the granite without having to remove their drill? It is inconceivable to me that they could have achieved such a depth with a hand-operated drill without the frequent withdrawal of the drill to clean out the hole or without their making provisions for the removal of the waste while the drill was still cutting. By frequently withdrawing the drill, however, they would have been able to expose their error and notice the direction their drill was taking before it had cut a .200-inch gouge into the side of the coffer and before it had reached a depth of eight or nine inches. The same situation applies as easily with the drill as with the saw—a high-speed operation made an error before the operators had time to correct it.

Although the ancient Egyptians are not given credit for having the wheel, the fact is that archaeological evidence, when evaluated with a machinist's eye, proves that they not only had the wheel, but they used it in very sophisticated ways. The evidence of lathe work is markedly distinct on some of the artifacts housed in the Cairo Museum, as well as those that were studied by Petrie. And if the Egyptians indeed used a lathe, then they had developed the wheel, for the products turned on a lathe, being circular, have the elements of being wheels—in fact, the wheels that are used on locomotives are turned on lathes. So although the lathes used by the ancient Eygptians have long disappeared, Petrie was very clear that they had existed when he identified the marks of true lathe turning on two pieces of diorite in his collection. It is true that intricate objects can be created without the aid of machinery

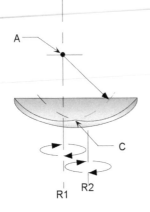

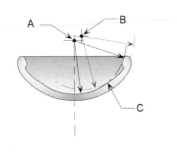

R1 and R2 indicate the rotational axes of the bowl. "A" is the pivot point of the tool, and "C" is the cusp where the radii intersect.

This is the original pivot point of the tool "A". The secondary pivot point of the tool "B" creates the bowl's lip and the cusp "C" where the radii intersect.

How the two bowl shards discovered by Petrie would have been machined.

FIGURE 19. *Petrie's Bowl Shards*

simply by rubbing the material with an abrasive such as sand or using a piece of bone or wood to apply pressure. The relics Petrie was looking at, however, in his words, "could not be produced by any grinding or rubbing process which pressed on the surface."[6]

The simple rock bowl shards Petrie was studying would hardly be considered remarkable to the inexperienced eye. However, Petrie, devoting as much care to the observation of this artifact (a) as he did to others, found that the spherical concave radius forming the dish had an unusual feel to it. Upon closer examination, he detected a sharp cusp where two radii intersected, indicating that the radii were cut on two separate axes of rotation (see Figure 19).

I have witnessed the same condition when a component has been removed from a lathe and then worked on again without being recentered properly. On examining other pieces from Giza, Petrie found another bowl shard that had the marks of true lathe turning (b). This time, though, instead of shifting the workpiece's axis of rotation, a second radius was cut by shifting the pivot point of the tool. With this radius, they machined just short of the perimeter of the dish, leaving a small lip. Again, a sharp cusp defined the intersection of the two radii.

While browsing through the Cairo Museum, I found evidence of lathe turning on a large scale. A sarcophagus lid had distinct lathe turning marks. The radius of the lid terminated with a blend radius at shoulders on both ends. The tool marks near these corner radii were the same as those I have observed when turning an object with an intermittent cut. The tool is deflected under pressure from the cut, and then relaxes when the section of cut is finished. When the workpiece comes round again to the tool, the initial pressure causes the tool to dig in. As the cut progresses, the amount of "dig in" is diminished. On the sarcophagus lid in the Cairo Museum, tool marks indicating these conditions were exactly where one would expect to find them (see Figure 20).

Egyptian artifacts representing tubular drilling are clearly the most astounding and conclusive evidence yet presented to indicate the extent to which machining knowledge and technology were practiced in prehistory. The ancient pyramid builders used a technique for drilling holes that is commonly known as "trepanning." This technique leaves a central core and is an efficient means of hole making. When making holes that did not go all the way through the material, the workers drilled to the desired depth and then broke the core out of the hole. Trepanning was evident not only in the holes that Petrie studied, but on the cores cast aside by the masons who had done the work. Regarding tool marks that left a spiral groove on a core taken out of a hole drilled into a piece of granite, Petrie wrote, "On the granite core, No. 7, the spiral of the cut sinks .1 inch in the circumference of 6 inches, or

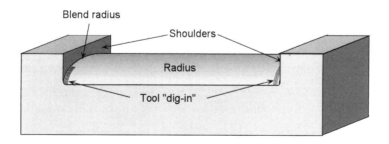

FIGURE 20. *Sarcophagus Lid in the Cairo Museum*

1 in 60, a rate of ploughing out of the quartz and feldspar which is astonishing."[7] After reading this, I had to agree with Petrie. This was an incredible feedrate (distance traveled per revolution of the drill) for drilling into any material, let alone granite. I was completely confounded as to how a drill could achieve this feedrate. Petrie was so astounded by these artifacts that he attempted to explain them at three different points in one chapter of his book.[8] To an engineer in the 1880s, what Petrie was looking at was an anomaly. The characteristics of the holes, the cores that came out of them, and the tool marks would be an impossibility according to any conventional theory of ancient Egyptian craftsmanship, even with the technology available in Petrie's day. Three distinct characteristics of the hole and core, as illustrated in Figure 21, make the artifacts extremely remarkable:

- A taper on both the hole and the core.
- A symmetrical helical groove following these tapers showing that the drill advanced into the granite at a feedrate of .10 inch per revolution of the drill.
- The confounding fact that the spiral groove cut deeper through the quartz than through the softer feldspar.

In conventional machining the reverse would be the case. In 1983 Donald Rahn of Rahn Granite Surface Plate Co. told me that diamond drills, rotating at nine hundred revolutions per minute, penetrate granite at the rate of one inch in five minutes. In 1996, Eric Leither of Tru-Stone Corp. told me that these parameters have not changed since then. The feedrate of modern drills, therefore, calculates to be .0002 inch per revolution, indicating that the ancient Egyptians drilled into granite with a feedrate that was five hundred times greater or deeper per revolution of the drill than modern drills! The other characteristics of the artifacts also pose a problem for modern drills. Somehow the Egyptians made a tapered hole with a spiral groove that was cut deeper through the harder constituent of the granite. If conventional machining methods cannot answer just one of these questions, how do we answer all three?

For those who may still believe in the "official" chronology of the historical development of metals, identifying copper as the metal the ancient

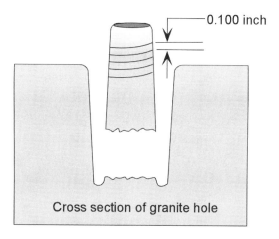

0.100 inch

Cross section of granite hole

Core removed from hole with lines
illustrates the feedrate of the drill.

FIGURE 21. *Petrie's Valley Temple Core & Hole*

Egyptians used for cutting granite is like saying that aluminum could be cut using a chisel fashioned out of butter. What follows is a more feasible and logical method, and it provides an answer to the question of techniques the ancient Egyptians may have used in all aspects of their work.

The fact that the feedrate evenly spirals along the length of the granite cores is quite remarkable considering the proposed method of cutting. The taper indicates an increase in the cutting surface area of the drill as it cut deeper, hence an increase in the resistance. A uniform feed under those conditions, using manpower, would be impossible. Petrie's theory of a ton or two of pressure being applied to a tubular drill consisting of bronze inset with jewels does not take into consideration that under several thousand pounds of pressure the jewels would undoubtedly work their way into the softer substance (the bronze), leaving the granite relatively unscathed after the attack. Nor does this method explain the groove being deeper through the quartz.

It should be noted that Petrie did not identify the means by which he inspected the core, whether he used metrology instruments, a microscope, or the naked eye. It also should be noted that Egyptologists do not universally accept his conclusions. In *Ancient Egyptian Materials and Industries,*

A. Lucas takes issue with Petrie's conclusion that the grooves were the result of fixed jewel points. He wrote:

In my opinion, to suppose the knowledge of cutting these gem stones to form teeth and of setting them in the metal in such a manner that they would bear the strain of hard use, and to do this at the early period assigned to them, would present greater difficulties than those explained by the assumption of their employment. But were there indeed teeth such as postulated by Petrie? The evidence to prove their presence is as follows.

> *(a) A cylindrical core of granite grooved round and round by a graving point, the grooves being continuous and forming a spiral, with in one part a single groove that may be traced five rotations round the core.*
> *(b) Part of a drill hole in diorite with seventeen equidistant grooves due to the successive rotation of the same cutting point.*
> *(c) Another piece of diorite with a series of grooves ploughed out to a depth of over one-hundredth of an inch at a single cut.*
> *(d) Other pieces of diorite showing the regular equidistant grooves of a saw.*
> *(e) Two pieces of diorite bowls with hieroglyphs incised with a very free-cutting point and neither scraped nor ground out.*

But if an abrasive powder had been used with soft copper saws and drills, it is highly probable that pieces of the abrasive would have been forced into the metal, where they might have remained for some time, and any such accidental and temporary teeth would have produced the same effect as intentional and permanent ones. . . .[9]

Lucas went on to speculate that the workers withdrew the tube-drill in order to remove waste and insert fresh grit into the hole, thereby creating the grooves. However, there are problems with this theory. It is doubtful that a simple tool that is being turned by hand would remain turning while the artisans draw it out of the hole. Likewise, placing the tool back into a clean hole with fresh grit would not require that the tool rotate until it was at the workface. There also is the question of the taper on both the hole and the core. Both would effectively provide clearance between the tool and the granite, thereby making sufficient contact to create the grooves impossible under these conditions.

In contrast, ultrasonic drilling fully explains how the holes and cores found in the Valley Temple at Giza could have been cut, and it is capable of creating all the details that Petrie and I puzzled over. Unfortunately for Petrie, ultrasonic drilling was unknown at the time he made his studies, so it is not surprising that he could not find satisfactory answers to his queries. In my opinion, the application of ultrasonic machining is the only method that completely satisfies logic, from a technical viewpoint, and explains all noted phenomena.

Ultrasonic machining is the oscillatory motion of a tool that chips away material, like a jackhammer chipping away at a piece of concrete pavement, except much faster and not as measurable in its reciprocation. The ultrasonic tool bit, vibrating at 19,000- to 25,000-cycles-per-second (hertz), has found unique application in the precision machining of odd-shaped holes in hard, brittle material such as hardened steels, carbides, ceramics, and semiconductors. An abrasive slurry or paste is used to accelerate the cutting action.[10]

The most significant detail of the drilled holes and cores studied by Petrie was that the groove was cut deeper through the quartz than through the feldspar. Quartz crystals are employed in the production of ultrasonic sound and, conversely, are responsive to the influence of vibration in the ultrasonic ranges and can be induced to vibrate at high frequency. When machining granite using ultrasonics, the harder material (quartz) would not necessarily offer more resistance, as it would during conventional machining practices. An ultrasonically vibrating tool bit would find numerous sympathetic partners, while cutting through granite, embedded right in the granite itself. Instead of resisting the cutting action, the quartz would be induced to respond and vibrate in sympathy with the high-frequency waves and amplify the abrasive action as the tool cut through it.

The tapering sides of the hole and the core are perfectly normal when we consider the basic requirement for all types of cutting tools. This requirement is that clearance be provided between the tool's nonmachining surfaces and the workpiece. Instead of having a straight tube, therefore, we would have a tube with a wall thickness that gradually became thinner along its length. The outside diameter becomes gradually smaller, creating clearance between the tool and the hole, and the inside diameter becomes larger, creating clearance between the tool and the central core. This would allow a free flow of abrasive slurry to reach the cutting area. By using a tube-drill of this design, the tapering of the sides of the hole and the core is explained. Typically this type of tube-drill is made of softer material than the abrasive, and the cutting edge would gradually wear away. The dimensions of the hole, therefore, would correspond to the dimensions of the tool at the cutting edge. As the tool became worn, the hole and the core would reflect this wear in the form of a taper (see Figure 22).

The requirement for advancing an ultrasonic tool into a workpiece is for the cutting edge of the tool to apply pressure to the workpiece as the vibratory motion of the tool does the actual cutting. This can be accomplished two ways: The tool can plunge straight down, or it can be screwed into the workpiece. We can explain the spiral groove if we select the latter method as the most likely one used. It should be made clear that the rotational speed of the drill is not a major factor in this cutting method; it is merely a means to advance the drill and apply pressure to the workpiece. Using a screw-and-nut method, the tube-drill could be efficiently advanced into the workpiece by turning it in a clockwise direction (see Figure 22). The screw would gradually thread through the nut, forcing the oscillating drill into the granite. It would be the ultrasonically induced motion of the drill that would do the cutting, not the drill-bit's rotation. The latter would be needed only to sustain a cutting action at the workface. By definition, the process is not a drilling process by conventional standards, but a grinding process in which abrasives are caused to impact the material in such a way that a controlled amount of material is removed.

The fact that there is a groove at all in the Valley Temple core may be explained several ways. An uneven flow of energy may have caused the tool to oscillate more on one side than the other, the tool may have been improp-

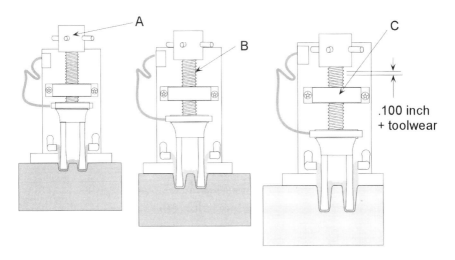

.100 inch
+ toolwear

Shows the progression of drilling in granite using ultrasonic
(vibratory) drill. The drill advances .100 inch plus toolwear
for every rotation of the handle (A).

Enlarged Cross Section of the Drill

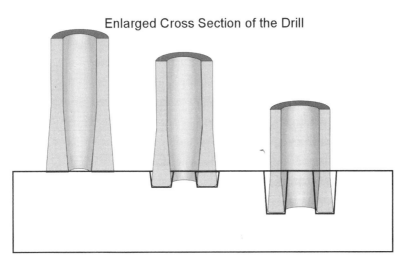

The abrasive slurry wears the tool as well as the granite.
The tool length diminishes as the cut deepens, resulting
in a taper on the core and the hole.

FIGURE 22. *Ultrasonic Drilling of Granite*

erly mounted, or a buildup of abrasive on one side of the tool may have cut the groove as the tool spiraled into the granite.

Another method by which the grooves could have been created was through the use of a spinning trepanning tool that had been mounted off-center to its rotational axis. Clyde Treadwell of Sonic Mill Inc. explained to me that when an off-centered drill rotated into the granite, it would gradually be forced into alignment with the rotational axis of the drilling machine's axis. The grooves, he claimed, could be created as the drill was rapidly withdrawn from the hole.

If Treadwell's theory is the correct one, it still requires a level of technology that is far more developed and sophisticated than what the ancient pyramid builders are given credit for. This method may be a valid alternative to the theory of ultrasonic machining, even though ultrasonics resolves all the unanswered questions where other theories have fallen short. When we search for a single method that provides an answer for all the data, we find that neither primitive nor most conventional machining methods provide that answer; consequently, we are forced to consider methods that are cutting-edge technologies even in our own time.

It goes without saying that further studies need to be made. One way to decide between opposing theories is to replicate the cores using the advanced machining methods I propose and the more primitive methods proposed by some Egyptologists. Following such a replication, the cores can be compared using metrology equipment and a scanning electron microscope in order to detect the microscopic changes in the structure of the granite that can result from the pressure and heat exerted or created by the tool. I doubt many Egyptologists share my conclusions regarding the pyramid builders' drilling methods, so it would be beneficial to perform these tests in order to prove conclusively the most likely method the builders used for cutting stone.

As this book was being prepared for publication, I received an unexpected e-mail from *NOVA*'s stonemason, Roger Hopkins, who had read my article about ancient Egyptian technology on the Internet. He wrote:

Dear Chris,

   You are a voice in the wilderness. I just finished reading your article about stone working techniques in ancient Egypt. I am a stone-

mason by trade and in 1991 the PBS series NOVA invited me to go to Egypt to experiment with building a pyramid; I quickly got bored with working the soft limestone and started to ponder the granite work. Here in Massachusetts, my specialty is working in granite (see my web page: http://tiac.net/users/rhopkins).

When I was asked by the Egyptologists how the ancients could have produced this work with mere copper tools, I told them they were crazy and that they were using at least state-of-art techniques. [At] first glance I tend to agree with you about the ultrasonic core hole drilling. I do enough core hole drilling to know that the embedded scrape marks would not be the result of ordinary core drilling. . . . I would love to explore this technique further with you and perhaps do a presentation in our next film about Egypt. . . .

Sincerely,

Roger Hopkins

In my subsequent communications with Hopkins, I found him to be very honest and straightforward regarding the techniques used by the ancients. His account of the building of the *NOVA* pyramid was much the same as that reported by Mark Lehner. He asked my permission to pursue the ultrasonic drilling aspects, as it was my idea, and I told him the more the merrier. The more people who are looking into how the ancient Egyptians accomplished their prodigious feats the better chance we have of determining the truth. Moreover, like any good businessman, Hopkins sees the potential for applying this technology in his own work.

My last e-mail from Hopkins informed me that he had contacted people at the Massachusetts Institute of Technology about pursuing this theory, had received promising feedback, and he would keep me informed. So the chapter on the ultrasonic drilling of Egyptian granite is at an end, even as this theory faces a new beginning.

Chapter Five

AMAZING DISCOVERY AT GIZA

 n February 1995, I joined Graham Hancock and Robert Bauval in Cairo to participate in a documentary. While there, I came across and measured some artifacts produced by the ancient pyramid builders that prove beyond a shadow of a doubt that highly advanced and sophisticated tools and methods were employed by this ancient civilization. Two of the artifacts in question are well-known; another is not, but it is more accessible, since it is lying out in the open, partly buried in the sand of the Giza Plateau. For this trip to Egypt I had taken along some instruments with which I had planned to inspect features I had identified during my 1986 trip. The instruments were:

- A "parallel"—a flat ground piece of steel about six-inches long and one-quarter-inch thick. The edges are ground flat within .0002 inch.
- An Interapid indicator (known as a clock gauge by my British compatriots).
- A wire contour gauge—a device once used, before the advent of computer numerical controlled machining, by die-makers to form around shapes.
- Hard forming wax.

I had taken along the contour gauge to check the inside of the mouth of the Southern Shaft inside the King's Chamber, for reasons to be discussed in a forthcoming chapter. Unfortunately, I found out after getting there that things had changed since my last visit. In 1993, a fan was installed inside this opening and, therefore, it was inaccessible to me and I was unable to check it. I had taken the parallel for quick checking of the surface of granite artifacts to determine their precision. The indicator was to be attached to the parallel for further inspection of suitable artifacts. Though the indicator did not survive the rigors of international travel, the instruments with which I was left were adequate for me to form a conclusion about the precision to which the ancient Egyptians were working.

Finding the King's Chamber in the Great Pyramid crowded with tourists, and not having the access I wanted to the Southern Shaft, I headed over to the Second Pyramid to inspect the "sarcophagus" there. Petrie had remarked that this granite box, like the one inside the King's Chamber, would had to have been installed in the bedrock chamber from above, before the chamber was roofed over and the pyramid finished, as it was too large to fit through the entrance passage. He supported his conclusion by pointing out that the bedrock chamber had gabled limestone beams that were put in place after the box was installed. Petrie's measurements of the passage were 41.08 to 41.62 inches wide by 47.13 to 47.44 inches high, and his dimensions of the box were 103.68 inches outside length, 41.97 inches outside width, 38.12 inches outside height; 84.73 inches inside length, 26.69 inches inside width, and 29.59 inches inside depth.[1] I.E.S. Edwards gave the angle of the entrance passage as 25°55'.[2] Petrie may have been correct in his assumptions, depending on how the smaller sloping passage is vertically oriented with the larger horizontal passage. Petrie was comparing the width of the box to the width of the passage, and obviously it will not fit. However, the box *will* fit into the smaller entrance passage if it is turned on its side. The only question not answered is whether there is enough room for it to tilt where the sloping passage meets the horizontal passage. It is unfortunate these questions were not on my mind at the time I was inside the pyramid, but my mission, at that time, involved other aspects of the ancient pyramid builders' work.

Crouching through the entrance passage and into the bedrock chamber, I climbed inside the box and—with a flashlight and the parallel—was astounded to find the surface on the inside of the box perfectly smooth and perfectly flat. Placing the edge of the parallel against the surface I shone my flashlight behind it. No light came through the interface. No matter where I moved the parallel—vertically, horizontally, sliding it along as one would a gauge on a precision surface plate—I could not detect any deviation from a perfectly flat surface.

A group of Spanish tourists found my activity extremely interesting, and they gathered around me as I animatedly demonstrated my discovery while exclaiming into my tape recorder, "Space-age precision!" The tour guides were becoming quite animated, too. I sensed that they probably did not think it was appropriate for a live foreigner to be where they believed a dead Egyptian should go, so I respectfully removed myself from the sarcophagus and continued my examination visually from the outside of the box.

There were more features of this artifact that I wanted to inspect, of course, but I did not have the freedom to do so. The corner radii on the inside appeared to be uniform all around, with no variation of precision of the surface to the tangency point. I was tempted to take a wax impression, but the hovering guides expecting bribes (baksheesh) inhibited this activity. (I was on a very tight budget.)

My mind was racing as I lowered myself into the narrow confines of the entrance shaft and climbed to the outside of the pyramid. The inside of a huge granite box had been finished off to an accuracy that modern manufacturers reserve for precision surface plates. How did the ancient Egyptians achieve this? And why did they do it? Why did they find that box so important that they would go to such trouble? It would be impossible to do that kind of work on the inside of an object by hand. Even with modern machinery it would be a very difficult and complicated task! Another point to consider was that the box, and the one in the King's Chamber inside the Great Pyramid, did not have to be made out of one piece if the only purpose it served was to house a dead body. There is evidence in the Cairo Museum proving that the ancient Egyptians also constructed sarcophagi out of five pieces and a lid. So why did they find it necessary to create each of these two boxes out of single blocks, which required the extra planning and effort to lower them into their chambers rather than drag them through the passages?

Petrie stated that the mean variation of the dimensions of the box in the Second Pyramid was .04 inch. Not knowing where the variation he measured was, I am not going to make any strong assertions except to say that it is possible to have an object with geometry that varies in length, width, and height and still maintain perfectly flat surfaces. Surface plates are ground and lapped to within .0001 to .0003 inch, depending on the grade of the specific surface plate; however, the thickness may vary more than the .04 inch that Petrie noted on that sarcophagus. A surface plate, though, is a single surface and would represent only one outside surface of a box. Moreover, the equipment used to rough and finish the inside of a box would be vastly different than that used on the outside. It would be much more problematic to grind and lap the inside of a box to the accuracy I had observed which would result in a precise and flat surface to the point where the flat surface meets the corner radius. The physical and technical problems associated with such a task are not easy to solve. One could use drills to rough the inside out, but when it comes to finishing a box of this size with an inside depth of 29.59 inches, while maintaining a corner radius of less than one-half inch, one would have to overcome some significant challenges.

While being extremely impressed with this artifact, I was even more impressed with other artifacts found at another site in the rock tunnels at the temple of Serapeum at Saqqara, the site of the Step Pyramid and Zoser's Tomb. I had followed Hancock and Bauval on their trip to this site for a filming on February 24, 1995. We were in the stifling atmosphere of the tunnels, where the dust kicked up by tourists lay heavily in the still air. These tunnels contain twenty-one huge granite and basalt boxes. Each box weighs an estimated sixty-five tons, and, together with the huge lid that sits on top of it, the total weight of each assembly is around one hundred tons. Just inside the entrance of the tunnels was an unfinished lid, and beyond this lid, barely fitting within the confines of one of the tunnels, was a granite box that also had been rough hewn.

The granite boxes were approximately 13-feet long, 7-1/2-feet wide, and 11-feet high. They were installed in "crypts" that were cut out of the limestone bedrock at staggered intervals along the tunnels. The floors of the crypts were about four feet below the tunnel floor, and the boxes were set into recesses in the center. Bauval had commented earlier about the engineering aspects of installing such huge boxes within a confined space where the last crypt was located near the end of the tunnel. With no room for the hundreds of slaves pulling on ropes to position these boxes, how were they moved into place?

While Hancock and Bauval were filming, I jumped down into a crypt and placed my parallel against the outside surface of the box. It was perfectly flat. I shone the flashlight and found no deviation from a perfectly flat surface. I clambered through a broken-out edge into the inside of another giant box and, again, I was astonished to find it astoundingly flat. I looked for errors and could not find any. I wished at that time that I had the proper equipment to scan the entire surface and ascertain the full scope of the work. Nonetheless, I was perfectly happy to use my flashlight and straightedge and stand in awe of this incredibly precise and incredibly huge artifact. Checking the lid and the surface on which it sat, I found them both to be perfectly flat. It occurred to me that this gave the manufacturers of this piece a perfect seal—two perfectly flat surfaces pressed together, with the weight of one pushing out the air between the two surfaces. The technical difficulties in finishing the inside of that piece made the sarcophagus in Khafra's Pyramid seem simple in comparison. Canadian researcher Robert McKenty was accompanying me at this time. He saw the significance of the discovery and was filming with his camera. At that moment I knew how Howard Carter must have felt when he discovered Tutankhamen's tomb.

The dust-filled atmosphere in the tunnels made breathing uncomfortable. I could only imagine what it would be like if I were a craftsman finishing off a piece of granite in that tunnel; regardless of the method I used, it would be unhealthy work. Surely it would have been better to finish the work in the open air? I was so astonished by this find that it did not occur to me until later that the builders of these relics, for some esoteric reason, intended for them to be ultra precise. They had gone to the trouble to take the unfinished product into the tunnel and finish it underground for a good reason. It is the logical thing to do if you require a high degree of precision in the piece that you are working. To finish it with such precision at a site that maintained a different atmosphere and a different temperature, such as in the open under the hot sun, would mean that when it was finally installed in the cool, cavelike temperatures of the tunnel, the workpiece would lose precision. The granite would give up its heat, and in doing so change its shape through contraction. The solution then as now, of course, was to prepare precision objects in a location that had the same heat and humidity in which they were going to be housed.

This discovery, and the realization of its critical importance to the artisans that built it, went beyond my wildest dreams of discoveries to be made in Egypt. For a man of my inclination, this was better than King Tut's tomb. The Egyptians' intentions with respect to precision are perfectly clear, but to what end? Further studies of these artifacts should include thorough mapping and inspection with the following tools:

- *A laser for checking surface flatness*—typically used for aligning precision machine beds.
- *An ultrasonic thickness gauge*—to check the thickness of the walls to determine their consistency to uniform thickness.
- *An optical flat with monochromatic light source*—to determine if the surfaces really are finished to optical precision.

I have contacted four precision granite manufacturers in the United States and not one can do this kind of work. In correspondence with Eric Leither of Tru-Stone Corp., I discussed the technical feasibility of creating several Egyptian artifacts, including the giant granite boxes found in the bedrock tunnels at the temple of Serapeum at Saqqara (see Figure 23). He responded as follows:

Dear Christopher,

First I would like to thank you for providing me with all the fascinating information. Most people never get the opportunity to take part in something like this. You mentioned to me that the box was derived from one solid block of granite. A piece of granite of that size is estimated to weigh 200,000 pounds if it was Sierra White granite which weighs approximately 175 lb. per cubic foot. If a piece of that size was available, the cost would be enormous. Just the raw piece of rock would cost somewhere in the area of $115,000.00. This price does not include cutting the block to size or any freight charges. The next obvious problem would be the transportation. There would be many special permits issued by the D.O.T. and would cost thousands of dollars. From the information that I gathered from your fax, the Egyptians moved this piece of granite nearly 500 miles. That is an incredible achievement for a society that existed hundreds of years ago.

Eric went on to say that his company did not have the equipment or

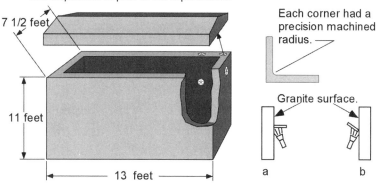

The lid was pushed to the back of the box, allowing the inspection of part of the top surface.

7 1/2 feet

Each corner had a precision machined radius.

11 feet

Granite surface.

13 feet

a   b

✪ Identifies areas that were inspected with a 6-inch-long flat ground steel straight-edge. There was no deviation from a flat surface.

No light from a flashlight leaked between the ground steel and the granite (a).

When checked with corner of the steel, (b) there were slivers of light. This would be the variation in the corner of the steel that was deburred using a file and was not as accurate as the edge.

FIGURE 23. *Granite Box in the Rock Tunnels at Saqqara*

capabilities to produce the boxes in this manner. He said that they would create the boxes in five pieces, ship them to the customer, and bolt them together on site.

Another artifact I inspected was a piece of granite that I, quite literally, stumbled across while strolling around the Giza Plateau later that day. I concluded, after doing a preliminary check of this piece, that the ancient pyramid builders had to have used a machine with three axes of movement (X-Y-Z) to guide the tool in three-dimensional space to create it. This artifact is very precise, even though it is a complex, contoured shape. Flat surfaces, having a simple geometry, can justifiably be explained as having been created by simple methods. This piece, though, because of its shape, drives us beyond the question, "What tools were used to cut it?" to a more far-reaching question, "What guided the cutting tool?" To properly address this question and be comfortable with the answer, it is helpful for us to have a working knowledge of contour machining.

THE GIZA POWER PLANT

Many of the artifacts that modern civilization creates would be impossible to produce using simple handwork. We are surrounded by artifacts that are the result of men and women employing their minds to create tools that overcome physical limitations. We have developed machine tools to create the dies that produce the aesthetic contours on the cars that we drive, the radios we listen to, and the appliances we use. To create the dies to produce these items, a cutting tool has to accurately follow a predetermined contoured path in three dimensions. The development of computer software has allowed some applications to move in three dimensions, while simultaneously using three or more axes of movement. The Egyptian artifact that I was looking at required a minimum of three axes of motion to machine it. When the machine-tool industry was relatively young, techniques were employed where the final shape was finished by hand, using templates as a guide. Today, with the use of precision computer numerical control machines, there is little call for handwork. A little polishing to remove unwanted tool marks may be the only handwork required. To know that an artifact has been produced on such a machine, therefore, one would expect to see a precise surface with indications of tool marks that show the path of the tool. This is what I found on the Giza Plateau, lying out in the open south of the Great Pyramid about one hundred yards east of Khafre's Pyramid (see Figure 24).

There are so many rocks of all shapes and sizes lying around this area that to the untrained eye these could easily be overlooked. To a trained eye, they may attract some cursory attention and a brief muse. I was fortunate that they both caught my attention and that I had some tools with which to inspect them. There were two pieces lying close together, one larger than the other. They had originally been one piece and had been broken. I found I needed every tool I had brought with me to inspect it. I was most interested in the accuracy of the contour and its symmetry.

What we have is an object that, three dimensionally as one piece, could be compared in shape to a small sofa. The seat is a contour that blends into the walls of the arms and the back. I checked the contour using the profile gauge along three axes of its length, starting at the blend radius near the back and ending near the tangency point, which blended smoothly where the contour radius meets the front. The wire radius gauge was not the best way to determine the accuracy of this piece. When adjusting the wires at one

1. The wax impression is taken.

2. The radius is verified.

3. The radius is verified at another location.

4. This is one end of the block that is broken off from the larger piece.

5. The radius is verified on the smaller separated block.

6. The accuracy is checked along the axial length of the block.

7. The undercut is a common engineering design feature that allows the use of a larger tool.

8. The contour is checked using a wire gauge.

9. The contour is verified at another location along the same axis.

FIGURE 24. *Contoured Block of Granite*

position on the block and moving to another position, the gauge could be reseated on the contour, but questions could be raised as to whether the hand that positioned it compensated for some inaccuracy in the contour. However, placing the parallel at several axial positions on the contour, I found the surface of this artifact to be extremely precise. At one point near a crack in the piece, there was light showing through, but the rest of the piece allowed very little to show.

During this time, I had attracted quite a crowd. It is difficult to traverse the Giza Plateau at the best of times without getting attention from the camel drivers, donkey riders, and purveyors of trinkets. It was not long after I had pulled the tools out of my backpack that I had two willing helpers, Mohammed and Mustapha, who were not at all interested in compensation. At least that is what they told me, although I can say that I literally lost my shirt on that adventure. I had washed sand and dirt out of the corner of the larger block and used a white T-shirt that I was carrying in my backpack to wipe the corner out so I could get an impression of it with forming wax. Mustapha talked me into giving him the shirt before I left, and I was so inspired by what I had found I tossed it to him. My other helper, Mohammed, held the wire gauge at different points along the contour while I took photographs of it. I then took the forming wax and heated it with a match—kindly provided by the Movenpick hotel—then pressed it into the corner blend radius. I shaved off the splayed part and positioned it at different points around. Mohammed held the wax still while I took photographs. By this time there were an old camel driver and a policeman on a horse looking on.

What I discovered with the wax was a uniform radius, tangential with the contour, the back, and the side wall. When I returned to the United States, I measured the wax using a radius gauge and found that it was a true radius measuring 7/16 inch. This, I believe, is a significant finding, but it was not the only one. The side (arm) blend radius, I found, has a design feature that is a common engineering practice today. The ancient machinists had cut a relief at the corner, a technique that modern engineers use to allow a mating part with a small radius to match or butt up against a surface with a larger blend radius. This feature provides for a more efficient machining operation because it allows the use of a cutting tool with a large diameter and, therefore, a large radius. This means that the tool has greater rigidity, and more material can be removed when making a cut.

I believe there is more, much more, that can be gleaned from ancient artifacts using these and other methods of study. I am certain that the Cairo Museum contains many artifacts that when properly analyzed will lead to the same conclusion that I have drawn from this piece—modern craftspeople and the ancient Egyptians have much in common in their use of the same kinds of machining techniques. The evidence, from granite artifacts at Giza and other locations, that ancient craftspeople used high-speed motorized machinery, and what we might call modern techniques in nonconventional machining such as ultrasonics, warrants serious study by qualified, open-minded people who can approach the subject without prejudice or preconceived notions.

The implications of such discoveries are tremendous in terms of a more thorough understanding of the level of technology employed by the ancient pyramid builders. We are not only presented with hard evidence that seems to have eluded us for decades and that provides support for the theory that the ancients were technically advanced. We are also provided with an opportunity to reanalyze history and the evolution—and devolution—of civilizations from a different perspective. But our understanding of *how* something was made opens up a different dimension when we then try to determine *why* it was made.

The precision in these artifacts is irrefutable. Even if we ignore the question of how they were produced, we are still faced with the question of why such precision was needed. Revelation of new data invariably raises new questions. In this case it is understandable for skeptics to ask, "Where are the machines?" But machines are tools, and the question should be applied universally and can be asked of anyone who believes other methods may have been used. The truth is that no tools have been found to explain *any* theory on how the pyramids were built or the granite boxes were cut. More than eighty pyramids have been discovered in Egypt, and the tools that built them have never been found. Even if we accepted the notion that copper tools are capable of producing these incredible artifacts, the few copper implements that have been uncovered do not represent the number of such tools that would have been used if every stonemason who is supposed to have worked on the pyramids at just the Giza site owned one or two. In the Great Pyramid alone there are an estimated 2,300,000 blocks of stone, both limestone and granite, weighing between two-and-one-half tons and seventy tons each. That is a mountain of evidence, and there are no tools surviving to explain even this one pyramid's creation.

The principle of Occam's razor, where the simplest means of manufacturing holds force until proven inadequate, has guided my attempt to erstand the pyramid builders' methods. In the theory proposed by Egyptologists, the basic foundation of this principle is lacking. The fact is the simplest methods do not satisfy the evidence, and Egyptologists have been reluctant to consider other, less simple methods. There is little doubt in my mind that Egyptologists have seriously underestimated the ancient builders' capabilities. But they have only to look at the precision of the artifacts and the evidence for the mastery of machining technologies, which have been recognized in recent years, to find some answers. It would also help to try and understand modern manufacturing at the shop floor level. Primitive methods, though simple to grasp intellectually, simply do not work in the field, and researchers would be well-served by gaining a better understanding of more sophisticated, ultra-precise methods.

One reference point for judging a civilization as advanced is to compare it with our current state of manufacturing evolution. Manufacturing is the physical manifestation of a society's scientific and engineering imagination and efforts. For over a hundred years industry has progressed exponentially. Since Petrie first made his critical observations of Egyptian artifacts between 1880 and 1882, our civilization has leapt forward technologically at breakneck speed. But, the development of machine-tools has been intrinsically linked with the availability of consumer goods and manufacturers' desire to find a customer. Most of our manufacturing development has been directed at providing the consumer with goods, which are created by artisans. Over a hundred years after Petrie, some artisans are still utterly astounded by the achievements of the ancient pyramid builders. They are astounded not so much by what they perceive a society is capable of creating using primitive tools, but rather by comparing these prehistoric artifacts with their own current level of expertise and technological advancement. To be objective, I must recognize that there are some artisans and engineers who resist revising their beliefs for the same reasons many Egyptologists do—they believe only "modern" societies are capable of sophisticated machining techniques. However, I would not be as bold in my assertions if I did not believe that the majority of my peers viewed the evidence with the same objectivity as I do and reached similar conclusions. I have presented this material to many engineers and artisans, and they are astonished at the evidence that is put before them.

To fully appreciate the value of this kind of research, we should keep in mind that the interpretation and understanding of a civilization's level of technology has predominately hinged on the preservation of written records. But for the majority of us, the nuts-and-bolts of our society do not always make interesting reading; in the same way, an ancient stone mural will more than likely have been cut to convey an ideological message rather than to preserve the information regarding the technique used to inscribe it. Moreover, the records of technology developed by our modern civilization rest in media that is vulnerable and could conceivably cease to exist in the event of a worldwide catastrophe, such as a nuclear war or another ice age. Our legacy will likely be read in the tangible remains of our society. Consequently, after several thousand years, someone looking back would most probably arrive at a more accurate interpretation of us and our society from our artisans' *methods* rather than an interpretation of our *language. The language of science and technology does not have the same freedom as does speech.* So even though the Egyptian tools and machines have not survived the thousands of years since their use, we have to assume, by objective analysis of the evidence for them left behind in the artifacts, that these tools did indeed exist.

There is much to be learned from our distant ancestors, if only we can open our minds and accept that another civilization from a distant epoch may have developed manufacturing techniques that are as great as or perhaps even greater than our own. As we assimilate new data and new views of old data, we are wise to heed the advice Petrie gave to an American who visited him during his research at Giza. The man expressed a feeling that he had been to a funeral after hearing Petrie's findings, which had evidently shattered some favorite pyramid theory he had at the time. Petrie said, "By all means let the old theories have a decent burial; though we should take care that in our haste none of the wounded ones are buried alive."[3]

With such a convincing collection of artifacts that prove the existence of precision machinery in ancient Egypt, the idea that the Great Pyramid was built by an advanced civilization that inhabited the Earth thousands of years ago becomes more admissible. I am not proposing that this civilization was more advanced technologically than ours on all levels, but it does appear that as far as masonry work and construction are concerned they were exceeding current capabilities and specifications. Making routine work

of precision machining huge pieces of extremely hard igneous rock is aston-
ishingly impressive.

Considered logically, the pyramid builders must have developed their
knowledge in the same manner any civilization would—reaching their state
of the art through technological progress over many years. As of this writing, there is considerable research being conducted by professionals throughout the world who are determined to find answers to the many unsolved
mysteries indicating that our planet has supported other advanced societies
in the distant past. Perhaps when this new knowledge and insight are assimilated, the history books will be rewritten and, if humankind is able to
learn from historical events, then perhaps the greatest lesson we can learn is
now being formulated for the benefit of future generations. New technology
and advances in the sciences are enabling us to take a closer look at the foundations upon which world history has been built, and these foundations
seem to be crumbling. It would be illogical, therefore, to dogmatically ad-
here to any theoretical point concerning ancient civilizations.

Such a revisioning occurred in 1986 when a French chemist named
Joseph Davidovits rocked the world with a startling new theory on pyramid
construction. Davidovits proposed that the blocks used to construct the
pyramids and temples in Egypt were actually cast in place by pouring
geopolymer materials into molds. In 1982, Davidovits analyzed limestone,
given to him by French Egyptologist Jean-Philippe Lauer, which was taken
from the Ascending Passage of the Great Pyramid and also the outer casing
stones of the pyramid of Teti. In his book *The Pyramids: An Enigma Solved*,
coauthored with Margie Morris, he reported:

> *X-ray chemical analysis detects bulk chemical composition. These tests*
> *undoubtedly show that Lauer's samples are man-made. The samples*
> *contain mineral elements highly uncommon in natural limestone,*
> *and these foreign minerals can take part in the production of*
> *geopolymeric binder.*
>
> *The sample from the Teti pyramid is lighter in density than the*
> *sample from Khufu's pyramid (the Great Pyramid). The Teti sample*
> *is weak and extremely weathered, and it lacks one of the minerals*
> *found in the sample from the Great Pyramid. The samples contain*

some phosphate minerals, one of which was identified as brushite,
which is thought to represent an organic material occurring in bird
droppings, bone, and teeth, but it would be rare to find brushite in
natural limestone.[4]

Davidovits' theory received worldwide attention, and I was challenged by several people to reconcile the theory that I was proposing with his. I have no difficulty reconciling my analysis of the cutting methods of the ancient pyramid builders with what Davidovits proposed. And I am sure he will see our individual efforts in the same light.

Davidovits cited *Pyramids and Temples of Gizeh,* in which Petrie devoted an entire chapter to the tool marks found on various artifacts made of both igneous and sedimentary rock. These artifacts were found both inside and outside the Great Pyramid. The tool marks on the stone tell us that they were cut, not poured. Nevertheless, this oversight should not entirely discredit Davidovits' findings. Construction technology today employs many techniques—cutting, forming, and pouring to name a few. Thus I believe it is shortsighted for me, or for anyone else, to discover one method of manufacture or construction and present it as the only method used by the pyramid builders.

Davidovits made a strong argument for his cast-in-place theory by pointing out the impossibility of the Egyptians having moved the huge monolithic blocks of stone that were used to build the pyramids. In most construction projects, if there is an option to do so, it does make sense to prepare a mold, or form, and pour the material, if the alternative is lifting and moving large masses weighing up to two hundred tons. Davidovits claimed that he had solved the problems associated with moving such huge stones with his cast-in-place theory. However, evidence that argues against the casting of igneous-type rock can be found in the rock tunnels at Saqqara. These are the giant granite and basalt boxes that weigh in at around eighty tons each. The existence of a roughed-out box and more than twenty finished boxes situated underground essentially disproves the argument that they were cast. We can speculate that when the craftspeople finished working the rough box, which is now wedged in one of the underground passageways, they would have had to move it into place without the benefit of hundreds of workers. That in and of itself is an impossibility. Furthermore, the very fact that this one box is rough cut belies the use of a casting method. If the Egyptians had cast these objects, they would not have chosen the characteristics of the roughed-out box for their mold. The product would be much closer to the finished dimensions of the other boxes, and more than likely the surfaces would be flatter than they actually are. These speculations do not mean that the ancient Egyptians did not use geopolymers. They simply mean that there may have been more than one method used to build the pyramids. To bring this whole issue into clearer perspective, perhaps we should now pause from evaluating the artifacts themselves and consider the work of an eccentric visionary who came before Davidovits, a man who also claimed he knew the secret of how the pyramids of Egypt were built—and succeeded in proving it.

Chapter Six

THE CORAL CASTLE MYSTERY

 hile the cutting techniques of the ancient pyramid builders have been an ongoing topic for debate, they have not received the same attention and controversy as the methods that were used to lift and transport cyclopean blocks of stone. Egyptologists and orthodox believers of primitive methods argue that the huge blocks were moved and positioned using only manpower, but experts in moving heavy weights using modern cranes throw doubt on their theory.

My company recently installed a hydraulic press that weighs sixty-five tons. In order to lift it and lower it through the roof, they had to bring in a special crane. The crane was brought to the site in pieces transported from eighty miles away over a period of five days. After fifteen semitrailer loads, the crane was finally assembled and ready for use. As the press was lowered into its specially prepared pit, I asked one of the riggers about the heaviest weight he had lifted. He claimed that it was a 110-ton nuclear power plant vessel. When I related to him the seventy- and two hundred-ton weights of the blocks of stone used inside the Great Pyramid and the Valley Temple, he expressed amazement and disbelief at the primitive methods Egyptologists claim were used.

For many of us to whom the Egyptologists' orthodox theory seems implausible, it is enough just to argue the issue from a logical standpoint. For others, the debate becomes more meaningful when a proposed alternate method is demonstrated and proven to be successful. For that proof we must turn to the one man in the world who, by demonstration, has supported the claim, "I know the secret of how the pyramids of Egypt were built!" That man was Edward Leedskalnin, an eccentric Latvian who immigrated to the United States and who is now deceased. But he left many intriguing clues that persuade us he may indeed have known such secrets.

The nine-ton Gate

Ed's celestial friends

Ed's greatest achievement -- a thirty-ton Rock

His and her moon seats

The entrance to Ed's workshop

The flywheel inside Ed's workshop

The fenced-off area is where Ed quarried his stone

FIGURE 25. *Assorted Photographs of Coral Castle*

Leedskalnin devised a means to single-handedly lift and maneuver blocks of coral weighing up to thirty tons. In Homestead, Florida, using his closely guarded secret, he was able to quarry and construct an entire complex of monolithic coral blocks in an arrangement that reflected his own unique character. On average, the weight of a single block used in the Coral

Castle was greater than those used to build the Great Pyramid. He labored for twenty-eight years to complete the work, which consisted of a total of 1,100 tons of rock. What was Leedskalnin's secret? Is it possible for a 5-foot tall, 110-pound man to accomplish such a feat without knowing techniques that are undiscovered to our mainstream contemporary understanding of physics and mechanics?

Leedskalnin was a student of the universe. Within his castle walls, he had a 22-ton obelisk, a 22-ton moon block, a 23-ton Jupiter block, a Saturn block, a 9-ton gate, a coral rocking chair that weighed 3 tons, and numerous other items. A huge 30-ton block, which he considered to be his major achievement, was crowned with a gable-shaped rock. Leedskalnin somehow single-handedly created and moved these massive objects without the benefit of cranes and other heavy machinery, a feat that astounds many engineers and technologists, who compare these achievements with those employed by workers handling similar weights in industry today.

Leedskalnin's castle was not always located in Homestead, Florida. He thought he had found his Shangri-la near Florida City and was happily working away on his rock garden until one night several thugs attacked him. Being a small man, he was an easy mark for these cowards, and he became a changed man after the trauma. Such was his concern that he became obsessed with moving his rock garden to a safer area. To assist him in the effort, he contracted with a local truck driver to haul his large rocks from Florida City to Homestead. As they prepared to load a 20-ton obelisk onto the truck, Leedskalnin asked the truck driver to leave him alone for a moment. Once out of sight, the truck driver heard a loud crash. Hurrying back to his truck, he was stopped in his tracks by the sight before him, hardly believing his eyes. He had returned just in time to see Leedskalnin dusting off his hands, the huge obelisk loaded and weighing down his flatbed.

Once in Homestead, the trucker was asked to leave his flatbed overnight and return in the morning. He was doubtful that Leedskalnin would be able to fulfill his promise that the obelisk would be off the truck and erected in the place he had set out for it. It's a good thing the truck driver did not bet money against Leedskalnin's ability to fulfill his word, because when he returned the following morning, Leedskalnin had moved the monolith into position, just as he had promised.

For his stupendous feats of construction engineering, Leedskalnin received attention not only from engineers and technologists, but from the United States Government, who paid him a visit, hoping to be enlightened. Leedskalnin received the officials gracefully, but they left none the wiser. In 1952, falling ill and on his last legs, Leedskalnin checked himself into the hospital and slipped away from this life, taking his secrets about moving massive objects with him.

If we assume that Leedskalnin and the ancient pyramid builders were using similar techniques, we must reevaluate the requirements for the man-hours necessary to construct the Great Pyramid. Estimates provided by Egyptologists for the number of workers that built the Great Pyramid range between 20,000 and 100,000. Based on the abilities of this one man, quarrying and erecting a total of 1,100 tons of rock over a time span of twenty-eight years, the 5,273,834 tons of stone built into the Great Pyramid could have been quarried and put in place by only 4,794 workers. If we figure in the efficiencies to be gained from working in teams and the division of labor, we can reduce the number of workers and/or shorten the time needed to complete the task. Let us not forget the estimate given by Merle Booker of the Indiana Limestone Institute for the delivery of enough limestone to build a Great Pyramid. Using the same criteria—with respect to size and quantity—as the ancient builders, but using modern equipment, Booker estimated that all thirty-three Indiana limestone quarries would have to triple their average output to produce the stone. His estimate did not factor in any equipment failures, labor disputes, or acts of God. He estimated that *twenty-seven years* after the order was placed, the last stone would have been delivered! These numbers put Leedskalnin's accomplishments in their proper perspective.

I first visited Coral Castle in 1982 while vacationing in Florida. It soon became clear to me that Leedskalnin's claim was accurate—he did indeed know some secrets, perhaps even the very ones used by the ancient Egyptians. I returned to Homestead again in April 1995 to refresh my memory and, specifically, to closely examine a device that, in 1992, fueled a discussion between an engineer colleague, Steven Defenbaugh, and myself. Our discussion resulted in a speculation as to the methods that Leedskalnin had used.

Leedskalnin took issue with modern science's understanding of nature. He flatly stated that scientists are wrong. His concept of nature was simple: All matter consists of individual magnets, and it is the movement of these magnets within materials and through space that produces measurable phenomena—that is, magnetism and electricity and so on.

Whether Leedskalnin was right or wrong in his assertions, from his simple premise he was able to devise a means to single-handedly elevate and maneuver large weights, which would be impossible using conventional methods. There is speculation that he was employing electromagnetism to eliminate or reduce the gravitational pull of the Earth. These speculations are entertained by some and scoffed at by others whose feet are firmly planted in the "real world."

While at Coral Castle, I commented to a lady standing in Leedskalnin's workshop that it was quite a feat he had performed, and asked if she had any idea how he had done it. Fixing me with a measured look, she said, "Through the application of physics and mechanics such work can be done." Somehow sensing my esoteric bent, she commented that Thor Heyerdahl had dispatched wild speculation about how the huge stone statues on Easter Island were put in place when he reenacted the work by carving, moving, and erecting one.

Being alone, and wanting a photograph taken of myself in Leedskalnin's workshop, I did not want to be argumentative. Smiling, I handed her the camera and did not point out that Heyerdahl, unlike Leedskalnin, had an ample supply of willing and healthy natives. They provided sufficient manpower to satisfy the physical requirements for conventionally moving such large weights, even on rollers, and cantilevering them into an upright position. Heyerdahl was an energetic man, but, using those methods, he could not have done it alone. Moreover, Heyerdahl merely demonstrated that the job could be done using one particular method. Anyone who has worked in manufacturing knows that there are many ways of doing things. To devise a means to perform a given work and present it as the only way that such work could be done gives little credit to those who either might know of a better way or might look for a better method—and succeed in finding one.

When analyzing ancient engineering feats, and being faced with explaining technically difficult tasks, Egyptologists and archaeologists typically throw in more time and more people using primitive, simple tools and manpower. Unlike those conventional arguments regarding ancient civilizations, in the case of Ed Leedskalnin we cannot impose the view that the work was done employing masses of people, for it is well-documented that Leedskalnin worked alone.

Egyptologists claim to "know" how the Great Pyramid was built. To prove it, stones no heavier than two-and-one-half tons were hefted into place using a gang of workers, straining on ropes. Leedskalnin claimed to "know" how the Great Pyramid was built, and to prove it he moved a thirty-ton and other monolithic blocks of coral to build his castle. It is too bad the cameras were not on Leedskalnin as they were on the *NOVA* experiment. I believe that Leedskalnin's feat is more illustrative of the pyramid builders' methods. While I enjoyed *This Old Pyramid*, I was not too impressed with the results. After a tremendous amount of effort using modern tools and equipment, the crew managed to move a few blocks into place using only manpower. After recently talking to Roger Hopkins, who was the mason in charge of the construction of the pyramid for the *NOVA* film, I have a lot more respect for the effort and knowledge that he put into it under extremely arduous conditions. Hopkins is very straightforward and an honest craftsman who specializes in working in granite. Like me, he is convinced that the ancients were using state-of-the-art equipment to perform this work.

But the equipment and techniques Leedskalnin used, I would suggest, go beyond what we know as state of the art. What technique did he use? Can we regain the knowledge he took with him to his grave? What follows is a speculation about the techniques Leedskalnin may have used. It follows his basic premise regarding the nature of electricity and magnetism and leads to a conclusion that, I believe, has some semblance of logic. This speculation followed some basic rules for brainstorming—those that follow and that might eventually reveal the secret should do the same. First, there is no such thing as a stupid idea, and, second, what we have been taught about the subject may not necessarily apply when seeking and, hopefully, finding a real solution.

A paradigm shift in my perception of "antigravity" occurred when my coworker, Steven Defenbaugh, and I were discussing the subject with Judd Peck, the CEO of the company for which we both work. Peck asked the simple question, "What is antigravity?" In an attempt to define it I had to say, "A means by which objects can be lifted, overcoming the gravitational pull of the Earth." As I spoke, it occurred to me that we were already applying antigravitational techniques in our everyday life. When we get out of bed in the morning, we employ antigravity just by standing up. An airplane, a rocket, a forklift truck, and an elevator are technologies devised to overcome the effects of gravity. Even a car rolling along on its wheels is an antigravity device. Without the wheels and a propulsion system, it would be just dead weight.

I realized that I had been laboring under the assumption that in order to create an antigravity device, gravity should be a known and fully understood phenomenon and that, through the application of technology, out-of-phase gravity waves would have to be created in such a manner as to neutralize it. As any physicist will tell you, the nature of gravity still eludes us, as does the ability to produce interference gravity waves.

So what if there is no such thing as gravity? What if the natural forces we already know about are sufficient to explain the noted phenomenon we have labeled as gravity? If, as Leedskalnin claimed, all matter consists of individual magnets, wouldn't the known properties of a magnet be sufficient? We know that like poles repel, and unlike poles attract. We also know that we can suspend one magnet above another as long as we do not allow either of them to flip over so that the opposite poles attract each other. Magnets seek to attract and, left to themselves, will align their opposite poles to each other. A mag-lev train is a good example of an antigravity device employing magnets.

The Earth, having the properties of a large magnet, generates streams of magnetic energy that follow lines of force. These lines of force have been noted for centuries (see Figure 26). If we assume, as Leedskalnin did, that all objects consist of individual magnets, we also can assume that an attraction exists between these objects due to the inherent nature of a magnet seeking to align its opposite pole to another. Perhaps Leedskalnin's means of working with the Earth's gravitational pull was nothing more complicated than devising a means by which the alignment of magnetic elements within his coral blocks was adjusted to face the streams of individual magnets he claimed are streaming from the Earth with a like repelling pole.

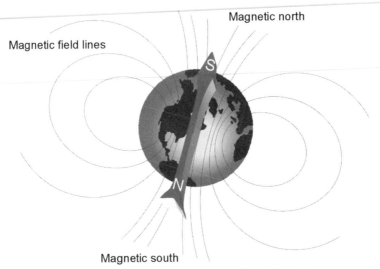

Magnetic north

Magnetic field lines

S

N

Magnetic south

The south pole of a magnetic compass needle is attracted
to the north pole of the Earth.

FIGURE 26. *Poles of Bar Magnet and Earth*

A well-known method for creating magnetism in an iron bar is to align
the bar with the Earth's magnetic field and strike the bar with a hammer.
This blow vibrates the atoms in the bar and allows them to be influenced by
the Earth's magnetic field. The result is that when the vibration stops, a
significant number of the atoms have aligned themselves within this magnetic field.

Was this the method that Leedskalnin was using? It is a simple concept, and when I observed the devices in Leedskalnin's workshop, I could
easily imagine the application of vibration and electromagnetism. His flywheel for creating electricity remains motionless, for the most part, until
inquisitive tourists like me come along and give it a spin. After giving it a
few revolutions, I realized that something was missing. The narrative I heard,
while browsing around the castle, described Leedskalnin as using this device to create electricity to power his electric lightbulbs. It was claimed that
Leedskalnin did not have electricity, but I could not imagine this device
being a useful and continuing source of power, considering Leedskalnin
used only his right arm to turn the wheel. On closer examination of the
piece, I found that the whole assembly was actually an old four-cylinder

crankcase. His flywheel was mounted on the front end of the crankshaft and consisted of bar magnets that were sandwiched between two plates—the upper plate being a ring gear. To give it weight and to solidify the entire assembly, Leedskalnin had encased the bar magnets with cement. It then occurred to me that the photo of Leedskalnin with his hand on the crank handle—which is attached to the end of the shaft—may not accurately represent his entire operation. It is possible that Leedskalnin was using the crank handle to start a reciprocating engine, now missing, which attached to one of the throws on the crankshaft. He would then be able to walk away and leave his flywheel running.

I was now mystified. I had developed a notion that the bars attached to the flywheel were actually being used to develop vibration in the piece Leedskalnin was trying to lift. This idea did not make sense considering the type of material, size, and weight of the entire assembly. The crankcase was firmly attached to the coral block in his workshop, and even if it was not attached, it would be quite a feat to keep moving it about. There was one factor I needed to check out, though, before I headed back to Illinois. I had tested the bar magnet with a pocketknife. The knife was attracted to each bar. I needed to know, conclusively, the arrangement of the poles in the wheel—to see, indeed, whether the assembly was capable of creating electricity.

I headed for the nearest strip mall to look for a hardware store so that I could buy a bar magnet. The first one had just what I needed—and for only $1.75. Feeling rather pleased, I returned to Coral Castle.

Once there, I headed back into Leedskalnin's workshop and put the magnet to the test. I held it a short distance away from the spokes of the flywheel while giving it a spin. Sure enough, I found out what I had come for. The magnet pushed and pulled in my grasp as the wheel rotated. Looking around the space, I gazed at a jumble of various devices, lying, hanging, and leaning about the room (see Figure 27). There were radio tuners, bottles with copper wire wrapped around them, spools of copper wire, and other various and sundry plastic and metal pieces that looked as if they had fallen out of an old radio set. Leedskalnin's workshop also contained chains, block and tackle, and other items that one might find lying around a junkyard. Some items were missing, though. Photographs of Leedskalnin at work show three tripods—made of telephone poles—that have boxes attached to the

FIGURE 27. *Ed Leedskalnin's Workshop*

top. These objects, however, are not to be found at Coral Castle. What is striking in the photograph is that the block of coral being moved is seen off to the side of the tripod. Perhaps Leedskalnin had moved the tripod after raising the block out of the bedrock. Another interesting observation is that the block and tackle that can be found inside his workshop is nowhere to be seen in this photograph. There are spools of copper wire in his workshop, and two wrappings of copper wire hang from nails in the wall. One was round copper and the other flat copper. In a narrative that visitors can hear at various recording stations around the compound, it is stated that at one time Leedskalnin had a grid of copper wire suspended in the air. Looking at the photograph again, one can see that there is a cable draped around the tripod and running down to the ground. Perhaps the arrangement of tripods was more related to the suspension of his copper grid than to the suspension of block and tackle.

If I were to try to replicate Leedskalnin's feat, I would begin with the premise that he was using his flywheel to generate a single-frequency tunable radio signal. The box at the top of the tripod would contain the radio receiver (there are several tuners in Leedskalnin's workshop), and the cable coming from the box would be attached to a speaker that emitted sound to vibrate the coral rock at its resonant frequency. With the atoms in the coral vibrating (like those in an iron bar), I would then attempt to flip their magnetic poles—which are naturally in an attraction orientation with the Earth—using an electromagnetic field.

Although today we stand in amazement before ancient megalithic sites that were built employing huge stones, if we had Leedskalnin's technique for lifting huge stones, it would make sense to us that the ancient masons might make their building blocks as large as possible. Very simply, it would be more economical to build in that manner. If we had a need to fill a five-foot cube, the energy and time required to cut smaller blocks would be much greater than what would be required to cut a large one.

I have no doubt that Leedskalnin told the truth when he said he knew the secrets of the ancient Egyptians. Unlike those who have sought publicity for their own inadequate, although politically correct, theories, he proved his theory through his actions. I believe, also, that we can rediscover his techniques and put them to use for the benefit of humankind. Edward Leedskalnin, right or wrong, had a little bit of a problem with trust—but this modus operandi was not unusual for a craftsperson of his day. Proprietary techniques without patent disclosure assure continued employment; therefore, it was perfectly normal that he would protect his secret from prying eyes that might steal and profit from it. I believe there are enough pieces of the puzzle in Leedskalnin's workshop to allow us to put them together and replicate his technique. It has been done once (sorry, twice!), and I am sure that it can be done again.[1]

Chapter Seven

ENDEAVORING TO EXPLAIN
THE ENIGMA

s we have seen, the evidence carved into the granite artifacts in Egypt clearly points to manufacturing methods that involved the use of machinery such as lathes, milling machines, ultrasonic drilling machines, and high-speed saws. They also possess attributes that cannot be produced without a system of measurement that is equal to the system of measure we use today. Their accuracy was not produced by chance, but is repeated over and over again.

After I assimilated the data regarding the ancient Egyptians' manufacturing precision and their possible and—in some instances—probable methods of machining, I suspected that to account for the level of technology that the pyramid builders seem to have achieved, they must have had an equally sophisticated energy system to support it. One of the pressing questions we raise when we discuss ancient ultrasonic drilling of granite is, "What did they use as a source of power?" A still more forceful inquiry regarding the use of electricity necessary to power ultrasonic drills or heavy machining equipment that may have been used to cut granite is, "Where are their power plants?" Obviously there are no structures from the ancient world that we can point to and identify as fission reactors, or turbine halls. And why should we have to? Isn't it a bit misguided of us to form an assumption that the ancient power plants were even remotely similar to ours?

Nevertheless, there may be some fundamental similarities between ancient and modern power supplies, in that the power plants in existence today are quite large and all need a supply of water for cooling and steam production. If such an advanced society existed in prehistory and if indeed they had an energy system, we could logically surmise that their power plants in all probability would be the largest construction projects they would attempt. It also may follow that, as the largest creations of the society, those power plants would stand a good chance of surviving a catastrophe and the erosion of the elements during the centuries that followed.

The pyramids easily meet these requirements. These geometric relics of the past, which have been studied, speculated about, and around which so much debate has centered, are located near a water supply, the Nile River, and, indeed, are the largest building projects that this ancient society completed. In light of all the evidence that suggests the existence of a highly advanced society utilizing electricity in prehistory, I began to seriously consider the possibility that *the pyramids were the power plants of the ancient Egyptians.*

Like just about every other student of the Egyptian pyramids, my attention was focused on the Great Pyramid, primarily because this is the one on which everybody else's attention had been focused, resulting in more research data being available for study. The reports of each successive researcher's discoveries inside the Great Pyramid are quite detailed, especially Petrie's. It is as though researchers became obsessed with reporting data, regardless of how insignificant it may have seemed.

Researchers have especially noted the Great Pyramid's geometric dimensions. Having worked with dimensions and angles all my working life—not just for the sake of dimensions themselves—their relevance in the Great Pyramid, while being important, was not of a primary concern to me. The dimensions, after all, are not the object, but a means to create the object. The area of study that I felt would reveal the true purpose of the Great Pyramid was a thorough examination of the inner chambers, passages, and every little detail that has been noted within them. While I was studying the inner chambers and passages of the Great Pyramid, I became convinced that I was looking at the prints for an extremely large machine, except this machine had been relieved of its inner components for some inexplicable reason. It is difficult to envision a machine that big, but with this basic premise I studied the drawings a little closer in order to obtain an understanding of how it might have operated.

The tremendous amount of masonry used in constructing this edifice suggested to me that there were things happening inside this pyramid that made such quantity necessary. Also, it seemed that the only logical reason an advanced civilization would have to build such a structure, at an obviously vast expenditure of time and energy, would be the same and only reason our civilization would need to duplicate it—to provide the population with some return on their investment. Energy would be such a return.

At the time the early explorers were crawling through the Great Pyramid, science and technology were at such a point that the basic foundations for explaining its true purpose had not yet been laid. At that time in our technological history, it would have been impossible for those researchers to entertain the thoughts recorded in this book; therefore, it is true to say that the science embodied in the Great Pyramid was lost to early explorers. But is this ancient science still lost?

I believe that the scientific foundation has been laid to attain that elusive lost science. In proposing my theory that the Great Pyramid is a power plant, I am not adamantly adhering to any one proposition. The possibilities may be numerous. However, the main facts are inescapable, for they were noted many years ago, and it would be impossible for an open-minded, logically thinking person to disregard them.

As we head toward the new millennium, the interest in ancient civilizations and the pyramids is gaining momentum. The idea that the Great Pyramid of Giza was built for the generation or as a source of energy is not new. Other authors have alluded to this idea and have made valuable contributions in the research of the Great Pyramid.

It is an exciting time we are living in. New information coming from Giza indicates that the theory that follows will find more evidence and proof to support it. Rudolph Gantenbrink's exploration in 1993 provided some of that proof, as I will shortly discuss. And testing in the Great Pyramid's King's Chamber by Tom Danley of the Schor expedition—though currently shrouded in secrecy imposed by a nondisclosure agreement—promises to reveal a tremendous amount of new and relevant information. Through the efforts of these researchers, and the groundwork laid by generations of others, enlightenment is about to dawn in the hazy world of Egyptology and we can be confident that the real truth about our distant ancestors in prehistory will soon come to light. The theory that follows, and the evidence I have gathered to support it, will, I hope, bring us one step further into the light of awareness about our species—where we were, where we are, and where we may be heading.

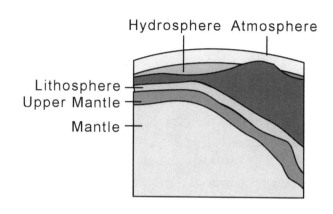

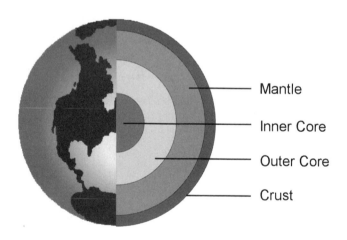

Figure 28. *The Earth's Layers*

Chapter Eight

THE GIZA POWER PLANT

*he Earth is a dynamic, energetic body that has supported civiliza-*
tion's demand for fuel for centuries. To date this demand has
predominantly been for energy in the form of fossil fuels. More
recently, scientific advances have allowed us to tap into the power
of the atom, and further research in this area promises greater
advances in the future. There is, however, another form of abundant energy in the Earth that in its most basic form has, for the most part,
been largely ignored as a potential source of usable energy. It usually gets
our attention when it builds up to a point of destruction (see Figure 28).

That energy is seismic, and it is the result of the Earth's plates being
driven by the constant agitation of the molten rock within the Earth. Most
earthquakes are the result of a shifting of these large, ridged blocks of rock,
or plates, that compose the Earth's surface. In a process called plate tecton-
ics, these plates are thrust against each other, away from each other, and side
to side. They do not slip freely, but build up energy over time and then slip
in a jerky fashion. Each jerk causes an earthquake because elastic energy
stored in the rock is suddenly released as seismic energy in the form of waves
that spread outward from the epicenter. The boundaries between these blocks
of rock are called faults, and it is at these points where the sudden shift occurs. (The San Andreas Fault is probably the most widely known fault in
North America.) Also contributing to strain within the Earth's crust is the
gravitational relationship between the Earth and the moon. The tides are
contained not only within the oceans of the world; the continents, too, are
in constant movement, rising and falling as much as a foot as the moon
orbits the Earth.

These earthquake vibrations produced in the Earth's outer layer, or crust,
can range from barely noticeable to catastrophically destructive. There are

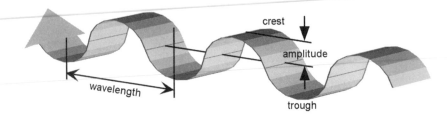

transverse S wave

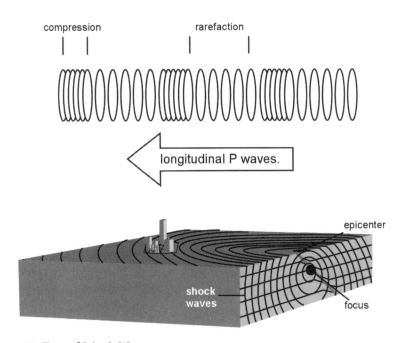

FIGURE 29. *Types of Seismic Waves*

six kinds of shock waves generated in the process (see Figure 29). Two types, known as body waves, travel through the Earth's interior, whereas the other four are surface waves. The movement of the rock distinguishes one kind of wave from another. Primary or compressional waves (P waves) send particles oscillating back and forth in the same direction as the waves are traveling. Secondary or transverse shear waves (S waves) oscillate perpendicular to their direction of travel. P waves always travel at higher velocities than S waves and are the first to be recorded by a seismograph.

Scientists believe that the Earth is analogous to a giant dynamo, with convection currents of charged molten metal circulating in the Earth's core. It is this flow of electric current in the core that generates its magnetic field (see Figure 30). Scientists have not yet seen the magnetism created by this flow of electricity as a potential source of energy because this field's force is relatively weak. Perhaps future technological innovations, of the kind that I suspect Edward Leedskalnin was using, will enable us to harness the Earth's electrical and magnetic energies. Who can say what doors new discoveries will open? The need for energy is ongoing, and as long as so much attention is focused on this subject, a decade should see many new innovations and changes.

With these considerations in mind, it would be helpful if we study another form of energy that is associated with a dynamo as the potential "raw material" for the production of power. Turn on any motor or generator and you can hear the energy at work: The motor/generator will hum as it revolves. This hum is associated with the energy itself and not so much the movements of the rotor through the air. This phenomenon is evident when a motor stalls while the power is still turned on. When too great a load is put

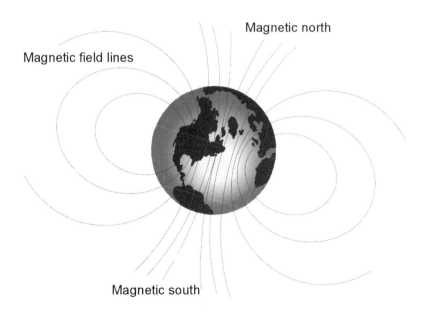

FIGURE 30. *Magnetic Fields of the Earth*

127

on a motor, and the motor stalls, the hum will become louder. The electrical and magnetic forces in the motor generate the sound waves. The Earth itself, as a giant dynamo, produces similar sound waves. The following is a brief explanation of this phenomenon:

> *Any local change in the density of an elastic medium can serve as a source for sound. This accounts for the great variety of acoustic sources because density changes may be produced in a great many ways, including mechanical, thermal, electrical, magnetic, and chemical actions. The most common sound waves are produced by the mechanical vibrations of solids, liquids, and gases. Solid vibrators include strings and rods, membranes and plates, shells (e.g., bells), as well as three-dimensional extended objects like the Earth itself. Liquid sources are not as common, but the turbulent flow of water or air provides an example. Gaseous sources include organ pipes, whistles, singing flames and explosions, as well as turbulent airflow.*[1]

There has been much discussion of late about the increasing frequency of the pulse of the Earth. There are speculations that the primary mode frequency appears to be gradually shifting higher, which therefore lends support to the belief that we are in for some significant Earth changes. Known as the Schumann Resonance, after German physicist W. O. Schumann who predicted the phenomenon between 1952 and 1957, these fundamental vibrations are the result of electrical activity between the Earth and its upper atmospheric layers. Collectively known as an electromagnetic "cavity," the elements that make it up are the Earth, the ionosphere, the troposphere, and the magnetosphere. The fundamental frequency of the vibrations is calculated to be 7.83 hertz, with overlaying frequencies of 14, 20, 26, 32, 37, and 43 hertz.[2]

Other researchers, however, contradict the idea that the Schumann Resonance is quickening because its frequency is related to the physical dimensions of the planet and the dimensional and electrical relationship between the surface of the Earth and the outer atmospheric layers. To increase the frequency would necessitate either a drastic change in the dimensions of the planet or the relocation of these outer layers to many miles within the

Earth. Like a guitar string that has a fixed length, or boundary, within which it will vibrate in response to the input of energy, the Schumann Resonance is the result of electrical activity within the boundaries made up of the surface of the Earth and the outer atmospheric layers. The tension, or resistance, to the energy in a guitar string can be variable, but in the atmosphere it is fixed at around 200 ohms.

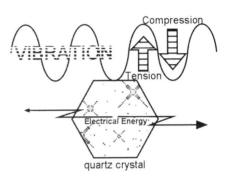

Vibration alternately compresses the crystal, producing electrical output.

FIGURE 31. *The Piezoelectric Effect*

The Earth's energy includes mechanical, thermal, electrical, magnetic, nuclear, and chemical action, each a source for sound. It would follow, therefore, that the energy at work in the Earth would generate sound waves that would be related to the particular vibration of the energy creating it and the material through which it passes. The audible hum of an electric motor—operating at 3,600 rpm—would fall well below the level of human hearing if it were to slow down to one revolution every twenty-four hours, as in the case of the Earth. What goes unnoticed as we go about our daily lives is our planet's inaudible fundamental pulse, or rhythm.

On the other end of the scale, any electrical stimulation within the Earth of piezoelectrical materials—such as quartz—would generate sound waves above the range of human hearing (see Figure 31). Materials undergoing stress within the Earth can emit bursts of ultrasonic radiation. Materials undergoing plastic deformation emit a signal of lower amplitude than when the deformation is such as to produce cracks. Ball lightning has been speculated to be gas ionized by electricity from quartz-bearing rock, such as granite, that is subject to stress.

It is not surprising that any sound generated by the electrical, magnetic, thermal, mechanical, and chemical action of the Earth goes unnoticed. With the influence of the ambient noise that surrounds us and that we create in our daily lives, we have managed to tune out any Earth sounds that may reach our ears. The birds, insects, and rustling of winds in the trees fill the countryside air with sound, and the large cities literally hum with activity.

As electrical energy can create mechanical vibrations (perceived as sound by the human ear), so in turn can mechanical vibrations create electrical energy, such as the previously mentioned ball lightning. It could be theorized, therefore, that with the Earth being a source for mechanical vibration, or sound, and the vibrations being of a usable amplitude and frequency, then the Earth's vibrations could be a source of energy that we could tap into. Moreover, if we were to discover that a structure with a certain shape, such as a pyramid, was able to effectively act as a resonator for the vibrations coming from within the Earth, then we would have a reliable and inexpensive source of energy.

So let us look at the Great Pyramid and its relationship to the Earth. Some incredible data have been recorded concerning the Great Pyramid that give us a clear insight into the builders' need to build a precise and close association with our planet. It could be passed off as coincidence that the Great Pyramid is located at the center of Earth's landmass (see Figure 32), but other characteristics of this structure strongly emphasize a close relationship to the Earth that is too significant to be overlooked.

When John Taylor wrote of the Great Pyramid, "It was *to make a record of the measure of the Earth* that it was built,"[3] he was basing this conclusion

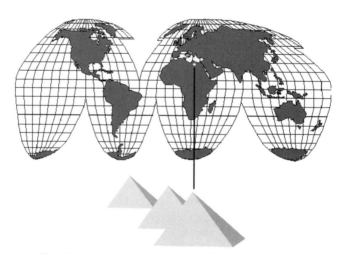

The Great Pyramid is at the center of the Earth's landmass.

FIGURE 32. *Center of Earth's Landmass*

on his evaluation of some astounding mathematical analyses, which had emerged through his research into the measurements of the Great Pyramid. He continued, "They knew the Earth was a sphere; and by observing the motion of the heavenly bodies over the Earth's surface, had ascertained its circumference, and were desirous of leaving behind them a record of the circumference as correct and imperishable as it was possible for them to construct."

It was the discovery of the knowledge of the transcendental number of pi (π) in the Great Pyramid that prompted Taylor to conclude that the perimeter of the Great Pyramid could be analogous to the circumference of the Earth at the equator. The height would represent the distance from the center of the Earth to the poles. Further studies of the dimensions of the Great Pyramid revealed surprising inferences regarding the knowledge of its builders. When searching for a unit that would fit the pyramid in whole numbers yet still retain the pi proportion, Taylor's answer of 366 base and 116.5 height suggested to him that the Egyptians may have divided the perimeter of the Great Pyramid into segments of the solar year. He also found the figure 366 when he divided the base of the pyramid by 25 inches. This suggested that the British inch was close to the Egyptian unit of measure, with 25 such units making one cubit.

It was later concluded that the Egyptian unit of measurement exceeded that British inch by .0011 inch, and Taylor found that this unit fit the Great Pyramid in multiples of 366. Even more astounding, geodetic research of the Earth established the Egyptian inch as an accurate unit of the dimensions of the polar radius. Peter Tompkins, in *Secrets of the Great Pyramid*, wrote: "To Taylor the inference was clear: the ancient Egyptians must have had a system of measurements based on the true spherical dimensions of the planet, which used a unit which was within a thousandth part of being equal to a British inch."[4] It was speculated that the British inch has lost a thousandth part after many generations of use.

Piazzi Smyth was a supporter of John Taylor and communicated with him frequently. Following Taylor's death in 1864, Smyth was able to confirm his calculations and also his correlation between the Great Pyramid and the Earth: ". . . and there appears to be further an even commensurability of a most marvelous order, between the weight of the whole Great Pyramid and the weight of our planet earth. *The Great Pyramid itself, found to be Harmo-niously Commensurable with the Earth, by Weight of the whole.*[5] Smyth cal-culated the weight of the Great Pyramid to be 5,273,834 pyramid tons and the weight of the Earth to be 5,273,000,000,000,000,000,000,000 pyramid tons. As such, he calculated it to be a 10^{15} integer of the Earth's weight.[6]

To review Taylor's findings:

- A pyramid inch is .001 inch larger than a British inch. There are 25 pyramid inches in a cubit and there were 365.24 cubits in the square base of the Great Pyramid.
- There are 365.24 days in a calendar year.
- One pyramid inch is equal in length to 1/500 millionth of the Earth's axis of rotation. This relationship suggests that not only were the build-ers of the Great Pyramid knowledgeable of the dimensions of the planet, they based their measurement system on them.

What else is unique about the Great Pyramid? Although it is a pyramid in shape, its geometry possesses an astounding approximation to the unique properties of a circle, or sphere. The pyramid's height is in relationship with the perimeter of its base as the radius of a circle is in relationship with its circumference. A perfectly constructed pyramid with an exact angle of 51°51' 14.3" has the value pi incorporated into its shape (see Table 2).

Pi (3.1415926) is an incommensurable number that, before calculators were invented, engineers used to round off to a three- or four-place decimal.

TABLE 2

	Petrie's Measurements	
Length of one side	9068.8 inches	755.733 feet
Perimeter (length x 4)	36275.2 inches	3022.93 feet
Height	5776.0 inches	481.33 feet
Angle based on above measurements	51°51'59"	
The Great Pyramid's approximation to pi based on Petrie's measurements	3.14017 (see Figure 33)	

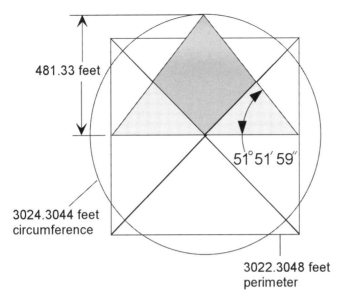

481.33 feet

51°51′59″

3024.3044 feet circumference

3022.3048 feet perimeter

Description	Value
Height of pyramid in feet	481.333 feet
Angle of pyramid at base	51°51′59″
Calculated length of one side	755.733 feet
Total perimeter	**3022.93 feet**
Radius of circle (R)	481.333 feet
Pi	3.141592654
Circumference (=R*2*Pi)	**3024.30 feet**

FIGURE 33. *The Pi Factor*

It is startling to read, therefore, that in 1883 Petrie published his painstaking measurements of the Great Pyramid and recorded the following: "On the whole, we probably cannot do better than take 51°52′ ± 2′ as the nearest approximation to the mean angle of the Pyramid, allowing some weight to the South side. The mean base being 9068.8 ± .5 inches, this yields a height of 5776.0 ± 7.0 inches."[7]

William Fix presented well-founded and objective data to support this claim: "We know that someone in very deep antiquity was aware of the size and shape of the earth with great precision. The three key measurements of the earth are incorporated in the dimensions of the Great Pyramid. The perimeter of the Pyramid equals a half minute of equatorial latitude. The perimeter of the sockets equals a half minute of equatorial longitude, or 1/43,200 of the earth's circumference. The height of the Pyramid including the platform, equals 1/43,200 of the earth's polar radius. . . . We do not know how they measured it, but that they did so is now an article of knowledge."[8]

There are volumes written on the measurements taken of the Great Pyramid, with each researcher producing slightly different results than the others. I am not going to attempt to argue for one data set or the other; however, I prefer Petrie's more realistic and "real-world" approach in reporting his data in that he provides a tolerance band within which his measurement may fall. Within the tolerance band he gave for the angle of the pyramid, we could have taken the perfect pi proportion, but it really is not necessary for the purposes of this book. What I have established by providing this data is that there is a distinct relationship between the Great Pyramid and the Earth. This is evidenced by the measurements of the pyramid and by its location.

When we question *why* there is a correlation between the Earth's dimensions and the Great Pyramid, we come up with three logical alternatives. One is that the builders wished to demonstrate their knowledge of the dimensions of the planet. They felt it necessary to encapsulate this knowledge in an indestructible structure so that future generations, thousands of years in the future, would know of their presence in the world and their knowledge of it.

The second possible answer could be that the Earth affected the function of the Great Pyramid. By incorporating the same basic measurements in the pyramid that were found on the planet, the efficiency of the pyramid was improved and, in effect, it could be a harmonic integer of the planet.

A third alternative may involve both the first and the second answers. The dimensions incorporated in the Great Pyramid may have been included to demonstrate the builders' knowledge or more importantly, to symbolize the relationship between the Great Pyramid's true purpose and the Earth itself. Perhaps the dimensions were not a critical requirement for the function of the pyramid, but were included to satisfy the aesthetic nature of the builders.

Considering the degree of practical consciousness evident in the Great Pyramid, I am inclined to accept a practical answer for noted phenomena and choose the second alternative because it recognizes a level of pragmatism that such a degree of consciousness would undoubtedly possess. The second alternative leaves no doubt that the builders of the Great Pyramid did not go to a vast amount of trouble simply to pass along their knowledge to some future generation. They used their energies for a more self-serving and timely purpose. That the builders of the Great Pyramid employed the dimensions of the Earth in their pyramid as a means to an end in achieving a specific result is easier to accept than the speculation that it was the result of some magnanimous gesture intended for future generations. Could our society afford to build—if we were able to—such a structure for this purpose?

Having established the relationship between the Great Pyramid and the Earth, it would be helpful if we remind ourselves that the measurements of an object are not the object itself but a means to create the object—or a means to an end. With the Great Pyramid the means are clearly evident. To what end they were used will become more obvious as we move along.

We know that the Earth is a vibrating dynamic body with tremendous forces that build up over time, forces that eventually result in a sudden release of a tremendous amount of energy. We might ask, therefore, "How can we tap into that energy?" Is there a way to draw the energy off over a period of time, thereby decreasing its intensity and possibly precluding the destructive forces of an earthquake? Science has shown us that it is possible, on a much smaller scale, for an object to draw mechanical energy from another vibrating object if both their vibrating frequencies are in harmony. But to draw mechanical energy out of the Earth would be a huge task. What would the requirements be for the object that we would use to accomplish such a thing?

Before we can answer these questions, we must refresh our minds regarding *resonance* and *harmonics,* for these are the natural phenomena that we would need to work with to accomplish such a task (see Figure 34). Resonance is the sympathetic vibration of one object with another. A piano provides a simple example of resonance. Press down one key, or several keys forming a chord, without actually striking the note, and then undamp the strings by pressing the loud pedal. Play the corresponding notes an octave higher, and the strings you have open on the lower octave will vibrate in sympathy. Hum into the piano in the same pitch and the strings will again respond. This transfer of energy is due to resonance. The transmission of energy and vibration go hand in hand. The strings of a musical instrument are induced to vibrate, and the energy reaches our ears in the form of sound waves.

When airborne sound forces mechanical vibrations in several piano strings that vibrate at different frequencies, the phenomenon known as harmonics is at work. Elements (strings) will absorb energy from a source more efficiently if they are of the same frequency. Multiples of the fundamental forcing frequency, known as harmonic frequencies, also will efficiently absorb this energy and vibrate at their natural resonance.

Resonance can probably best be described by a classic example of how this natural phenomenon can unleash an awesome and destructive power.

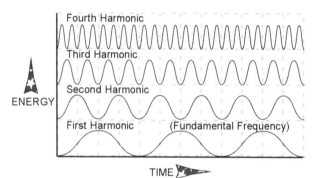

Elements will absorb energy and vibrate at their resonant frequency. If the resonant frequency of an element is a harmonic of a fundamental driving force, it will be induced to vibrate in sympathy.

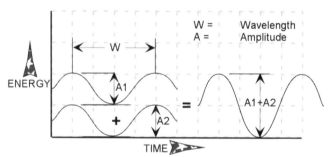

Two positive interference waves will reinforce each other resulting in an increase in amplitude (energy).

Frequency (Hertz) = Velocity (feet/second) / Wavelength

FIGURE 34. *Resonance and Harmonics*

This incident occurred on the morning of November 7, 1940, in the State of Washington. The Tacoma Narrows Bridge created a link between the Olympic Peninsula and the mainland. It had been open only for four months when tragedy struck. In gusts of wind at only forty-two miles per hour, the bridge began to oscillate, wildly swaying back and forth. The gusts of wind swept across the bridge at a frequency that matched the bridge's natural resonant frequency. As the gusts of wind continued, the bridge's torsional vibrations amplified to the point that the suspenders were torn away from their moorings and the bridge began to break up. Fortunately, it was closed to traffic in time, and no lives were lost. Nature's demolition of the Tacoma Narrows Bridge is given as a classic example of the destructive forces that can be induced in a structure that is subject to periodic influxes of energy. In this case, the energy was provided by the wind, which swept the bridge at the appropriate resonating frequency of the structure. With insufficient restraint, or damping, any vibrating structure may eventually be destroyed, as long as it is drawing energy from the source.

Another example of the potentially destructive force of resonance, and the measure taken to prevent it, is the instruction for soldiers to break step when marching across a bridge. Each step of an individual soldier acts as a force on the bridge. If the rest of the company joins this soldier in marching in unison across the bridge, the energy provided by that one step is amplified many times over, and the bridge will vibrate in time to the march. The pounding foot on the bridge is known as the forcing frequency. If the frequency of the marching feet happens to coincide with the natural frequency at which the bridge resonates, the absorption of energy will be maximized and the vibration of the bridge will become much greater, and it could potentially cause the bridge to collapse.

The *Encyclopedia Britannica* explains this phenomenon: "As has already been suggested, if the damping is very small, a vibrator will draw correspondingly large average power from the source, especially at resonance. If the damping becomes effectively zero, momentarily, or even negative, as can happen under certain peculiar circumstances, the power withdrawal may become so great as to lead to a runaway vibration that may destroy the vibrator."[9]

At the Army Corp of Engineers Research Laboratory in Champaign,

Illinois, there is a huge "shake table" that is used to test military equipment. The table can simulate the forces of an earthquake by generating both longitudinal and transverse waves. The vibration of the table can also be ramped up and down to any desired frequency. A demonstration of resonance using the shake table leaves one with a strong sense of the awesome power this unique phenomenon can unleash. To demonstrate the effect resonance can have on an object, the shake table is fitted with several long plastic tubes, around six inches in diameter and of various lengths. The tubes are clamped perpendicular to the tabletop. The table is then vibrated and slowly ramped up in frequency. The tubes slowly sway back and forth until the vibrations reach the resonant frequency of any one tube, and then that tube begins to oscillate more than the others. If that frequency is maintained for a period of time, that tube shakes wildly while the other tubes remain relatively stable. Before the tube shakes loose from its clamp, the operator raises the frequency and the oscillations of the one tube die down. The frequency of the table continues to increase until another tube begins to vibrate wildly. The process repeats itself across each of the tubes on the table. At its own unique resonant frequency, each tube will draw more energy from the source.

It is clear, then, that in order to draw mechanical vibrations and relieve the stresses that build up within the Earth, we would need an object that would respond sympathetically with the Earth's fundamental frequency. This object would need to be designed in such a way that its own resonant frequency was the same as, or a harmonic of, the Earth's. In this manner, energy transfer from the source would be at maximum load. In harmony with the Earth's vibrations, this object would have the potential to become a coupled oscillator. (A coupled oscillator is an object that is in harmonic resonance with another, usually larger, vibrating object. When set into motion, the coupled oscillator will draw energy from the source and vibrate in sympathy as long as the source continues to vibrate.)

Because the Earth constantly generates a broad spectrum of vibration, we could utilize vibration as a source of energy if we developed suitable technology. Naturally, any device that attracted greater amounts of this energy than is normally being radiated from the Earth would greatly improve the efficiency of the equipment. Because energy will inherently follow the path of least resistance, it follows that any device offering less resistance to this energy than the surrounding medium through which it passes would have a greater amount of energy channeled through it. Keeping all of this in mind and knowing that the Great Pyramid is a mathematical integer of the Earth, it may not be so outlandish to propose that the pyramid is capable of vibrating at a harmonic frequency of the Earth's fundamental frequency.

As it turns out, such is the case!

Acoustic data, scanty as it is, supports the theory that the Great Pyramid responds to vibrations from within the Earth. I wish I had been with NASA consultant and acoustic engineer Tom Danley when he conducted his acoustical analysis within the King's Chamber. I owe him a great deal because he performed a task that I have wanted to see transpire for twenty years. As of this writing, Danley is remaining very quiet about his research in Egypt because he is currently under a nondisclosure agreement with the Schor Foundation, a private research organization headed by industrialist Joseph Schor. The Schor Foundation funded Danley's research within the Great Pyramid in 1996 and, honoring an agreement with the Egyptian Department of Antiquities, they are keeping strict control of all the information they gathered there.

There was another member of this party, however, who is not keeping so quiet. Boris Said (pronounced Sa-eed) is a self-proclaimed documentarian and was a producer of *The Mysteries of the Sphinx* documentary. Said was under a nondisclosure agreement with Schor as well, but claims that Schor negated the agreement by not telling him that his permit to film at Giza had been withdrawn, or had expired, as he was still working on the plateau. In an interview on the Art Bell radio show, Said described Danley's experiments, which involved the use of large amplifiers, subwoofers, and accelerometers (an instrument designed to detect vibration) that were strategically placed in the King's Chamber and in each of the air spaces above the King's Chamber. Much of what Danley discovered remains with Danley, but what little Said has disclosed to promote a documentary video he produced is quite revealing:

> *Subsequent experiments conducted by Tom Danley in the King's Chamber of the Great Pyramid and in Chambers above the King's Chamber suggest that the pyramid was constructed with a sonic pur-*

pose. Danley identifies four resident[10] frequencies, or notes, that are enhanced by the structure of the pyramid, and by the materials used in its construction. The notes from [sic] an F Sharp chord, which according to ancient Egyptian texts were the harmonic of our planet. Moreover, Danley's tests show that these frequencies are present in the King's Chamber even when no sounds are being produced. They are there in frequencies that range from 16 Hertz down to 1/2 Hertz, well below the range of human hearing. According to Danley, these vibrations are caused by the wind blowing across the ends of the so-called shafts—in the same way as sounds are created when one blows across the top of a bottle.

Included in the program is a meeting with a Native American maker of sacred flutes from Oregon. His flutes, which are made to serenade Mother Earth, are tuned to the key of F Sharp![11]

Danley has come up with an interesting theory regarding the wind as the source of the infrasonic sound, but I wonder if he is really hiding something. Did his instruments reveal the source of the sound to be the Earth itself, but he is not allowed to tell? The reason I wonder is that fans have been installed within the mouths of the shafts Danley refers to; and in the west side of the passage leading into the King's Chamber is a tunnel bored to the northern shaft, which has been opened along its length for several feet, precluding any vibrations in the King's Chamber from being caused by the "Coke bottle" effect. Moreover, the fans were installed to *remove* excess heat and humidity, and are drawing air *from* the pyramid's chambers through the shafts to the outside. All of these are conditions that would make the "Coke bottle" effect extremely unlikely. Danley, as an acoustical engineer, must know this as well as I do. Therefore, I wonder if the most likely source for the infrasonic sound within the King's Chamber is the Earth itself.

I recently had the opportunity to confirm the acoustical phenomena of the King's Chamber in a unique and rather fortuitous way, although without Danley's instrumentation or expertise. On February 24, 1995, I paid the inspector of the Giza Plateau $100 to leave me inside the Great Pyramid after all the tourists had left and it was officially closed. It was Ramadan, a sacred time for Muslims, and tourist attractions were closing early. I had asked to be left alone in the Great Pyramid for thirty minutes with all the lights turned off. Mohammed, the inspector, thought I was going to meditate, and I did not correct his thinking as I negotiated the deal. My backpack was weighted with my water bottle, an essential item in Egypt, and some instruments I had brought along specifically to take some acoustical and electromagnetic frequency measurements. I was pressed for time, and the activities I had planned allowed no time for meditation. I had asked for the lights to be turned off because I did not want any background electrical noise to affect the digital frequency counter with which I was equipped. I had brought this along to measure radio frequencies that I believe can be generated by the resonant chamber inside the Great Pyramid. I also had brought along a tape recorder, which I turned on and placed upon a block of granite situated close to the granite coffer (I will call the coffer a "box" from this point forward) in the King's Chamber. Using this block as a work stage, I positioned my flashlight, the digital frequency counter, and a monochromatic tuner, which measures sound frequency and is used to tune musical instruments, as I heard the last batch of tourists making their way down the Ascending Passage to the outside.

When the noise faded away, I began to test the frequency of the granite box. I had read in a booklet by flautist Paul Horn (that accompanied his album *Inside the Great Pyramid*) that the granite box resonated at a frequency of 438 cycles per second (Hz).[12] Horn had used a Korg tuner, which is quite a bit more expensive than the Matrix tuner I was using, to perform his own acoustical tests inside the chamber. I thumped the side of the box with my fist. The tuner registered between 439 and 440 Hz. I loudly hummed that note to test the resonance of the King's Chamber. Sliding up the scale, I noted that the reverberation faded until I produced the same note one octave higher, and then the reverberation was even greater. It was then time to test for radio frequencies, and the lights were still not turned off. I was becoming concerned because I had to be out of there in fifteen minutes, so I crouched back through the passageway, stood at the top of the Grand Gallery, and yelled to the guards to turn off the lights. While there, I intoned the note I had hummed earlier, then scurried back into the King's Chamber with the intent of being by my flashlight before I was pitched into darkness. At last, the hum of the lights ended abruptly; I was left with only the light from my flashlight. My heart was pounding in my chest. I was sweating profusely. I had reached the moment of truth.

Not that day! To my chagrin, I realized that the hum of the lights was masking the whir of fans installed in the alleged "air channels." The guards had turned the lights off, but, as I discovered later, they could not turn off the fans without a key to the electrical box that controlled them. I "blessed" Rudolph Gantenbrink for such a fine engineering job, for he had installed those fans before he made his famous exploration of the Queen's Chamber shafts. Disappointed that I would not be able to use any data, I went ahead and took some readings anyway. I then packed my materials and hurried out of the King's Chamber, down the Grand Gallery, tortured my thighs again down the Ascending Passage, then walked quickly through Al Mamun's forced passage to the gate and out to fresh air.

Back in my hotel room, I rewound the tape, played it back, and discovered three very interesting acoustical phenomena. First, the tape had picked up overtones to the note I was humming in the King's Chamber. I had been unable to hear them while I was in the chamber because I was the source of the sound. This was a very exciting discovery, as it supported my theory on the King's Chamber. I tempered my enthusiasm, however, with the thought that it might be the equipment that was resonating at a higher frequency and generating this overtone. Mine was not a very expensive tape recorder, but still I believe the overtones were generated by the chamber complex itself (including the granite ceiling beams) because of the King's Chamber's true design and purpose. The second discovery revealed that when I was on the Great Step at the top of the Grand Gallery yelling to have the lights turned off and then humming the tone, the playback sounded as if I had never left the room. At that moment, except for a small passageway, there were thousands of tons of granite and limestone that separated me from the recorder, and my voice was being projected into a 28-foot-high and 157-foot-long expanse of space. The third revelation was that the footsteps and noises I made, while traversing the low passage to and from the Grand Gallery, reverberated in the King's Chamber and caused it to resonate at its natural frequency. This was noted on the Matrix monochromatic tuner that was turned on as I played back the tape. My footsteps and noises were registering at approximately 440 Hz. I particularly noted that the fluctuations of the tuner during this section of the playback were not as great as those I had observed inside the King's Chamber when I was humming the frequency.

On June 8, 1997, I was discussing these phenomena with Stephen Mehler, the director of research for the Kinneman Foundation. In February 1997 after appearing with Mehler on a television panel discussion in Santa Barbara, California, I had sent him an abstract of my theory regarding the Great Pyramid. He told me about an acoustics engineer named Robert Vawter who had done some studio analysis of a tape recording that he (Mehler) had made inside the King's Chamber. As he related it to me, those results were the same as mine. I was given Vawter's phone number and promptly gave him a call. Vawter confirmed what Mehler told me. He said that he had digitally processed the tape provided by Mehler, and he was able to isolate harmonic overtones of the intoned frequency. Vawter claimed that the King's Chamber was designed specifically as a resonant chamber in which sound of specific frequencies would resonate. He said that every dimensional feature of the chamber he had studied indicated the manifestation and form of harmonic resonance. He was in the process of compiling data to support his statement and will be publishing his findings in the future.

While I was discussing with Mehler my experience inside the King's Chamber and the noise of the tourists as they were leaving the pyramid, he related a report by Howard-Vyse, who claimed that he was in the King's Chamber and was able to hear a conversation taking place in the Subterranean Pit. Vawter confirmed the speculation that this is because the entire interior passageway was designed to maximize the throughput of sound. This same phenomenon is well known at St. Paul's Cathedral in London. The Whispering Gallery has surprised many a visitor, who hears a voice and turns around to see who is speaking, only to find no one there. The gallery is circular, with walls made of hard material. The sound of a person speaking in a low voice on one side of the gallery will be reflected around the circular wall, just grazing the hard surface, and can be heard on the other side of the gallery. The angle of incidence ensures total reflection of the sound.

With this experimental evidence available, and with what can be extrapolated from the dimensions and mass of the Great Pyramid, we have an object that fits the criteria established as necessary for an object to draw vibrations from the Earth. That object is the Great Pyramid of Giza! Here is the product of an ancient civilization empowered with the knowledge that as long as the moon continued to orbit the Earth, the special relationship that existed between the two assured the Egyptians of vast amounts of energy. The source of the energy is the Earth itself, in the form of seismic energy. The ancient Egyptians saw tremendous value in this form of energy and expended a considerable amount of effort to tap into it. The benefits they received may have been twofold: energy to fuel their civilization, and the ability to stabilize the Earth's crust by drawing off seismic energy over a period of time rather than allowing it to build up to destructive levels.

Covering a large land area, the Great Pyramid is, in fact, in harmonic resonance with the vibration of the Earth—a structure that could act as an acoustical horn for collecting, channeling, and/or focusing terrestrial vibration. We are led to consider, therefore, that energy associated with the pyramid shape is not drawn from the air or magically generated simply by the geometric form of a pyramid, but that *the pyramid acts as a receiver of energy from within the Earth itself.* It could be, also, that these infrasonic sound waves provide an explanation for the physical phenomena some people have felt when entering the Great Pyramid. The "pyramid energy" that has inspired countless numbers of people since the time of Napoleon may be the effects of infrasonic sound on the brain, which is said to resonate at around 6 hertz.

I experienced the phenomenon myself while in Egypt in 1986. After being inside the Great Pyramid for about an hour, I found myself in a rather uncomfortable situation. Sick and in immediate need of a bathroom, I really did not know if I was going to make it; but I rushed out of the King's Chamber, down the Grand Gallery and Ascending Passage, squeezing past tourists. Once outside I ran down the hill to the Mena House and headed straight for the bathroom. I made it just in time. The bathroom walls in the Mena House were constructed of Aswan granite. As I relaxed and closed my eyes, and without any external influence, the resonance of the King's Chamber filled my head. At the same time a pyramid shape began to glow in the center of my forehead. It was only after leaving the Mena House that this sensation faded. It could be that this phenomenon is only felt at certain times, according to the seismic activities within the Earth. I have not experienced it since then, although I have been inside the Great Pyramid several times since.

While infrasonic vibrations at around 6 hertz may influence the brain and produce various effects in humans, it seems that there must be other types of energy, or other frequencies, to explain phenomena that were noted to have occurred at the Great Pyramid more than one hundred years ago. Sir William Siemens, an Anglo-German engineer, metallurgist, and inventor, experienced a strange energy phenomenon at the Great Pyramid when an Arab guide called his attention to the fact that, while standing on the summit of the pyramid with hands outstretched, he could hear a sharp ringing noise. Raising his index finger, Siemens felt a prickling sensation. Later on, while drinking out of a wine bottle he had brought along, he experienced a slight electric shock. Feeling that some further observations were in order, Siemens then wrapped a moistened newspaper around the bottle, converting it into a Leyden jar. After he held it above his head for a while, this improvised Leyden jar became charged with electricity to such an extent that sparks began to fly. Reportedly, Siemens' Arab guides were not too happy with their tourist's experiment and accused him of practicing witchcraft. Peter Tompkins wrote, "One of the guides tried to seize Siemens' companion, but Siemens lowered the bottle towards him and gave the Arab such a jolt that he was knocked senseless to the ground. Recovering, the guide scrambled to his feet and took off down the Pyramid, crying loudly."[13]

M. Bovis, a Frenchman, visited the Great Pyramid and on passing through the King's Chamber he spotted some dead cats and other animals in a garbage can. Bovis noted that these animals did not have the usual putrid odor that is normally associated with decaying flesh and he became intrigued with this discovery, for the animals appeared to be dehydrated or mummified. Curious to find what conditions were creating this phenomenon, Bovis intuitively gave his attention to the actual shape of the pyramid. Upon his return to France, he constructed his own small pyramid, using a three-foot base and maintaining the precise 51°51' angle of the Great Pyramid. To his delight, he found that he could duplicate the mummification process he had observed in the pyramid and, taking his experiments further, he found that fruits and vegetables could be preserved also.

The phenomenon Bovis noted may indicate the presence of ultrasonic radiation within the pyramid. His claims, and those of other proponents of pyramid energy, seem to correlate, at first glance, with some applications that have been found for ultrasonic sound:

- *The aging of fermented beverages.* It has been noted that wine tastes smoother after being treated with pyramid energy. A modern technique in speeding up the aging process in wine is to irradiate the wine with ultrasonic sound. Perhaps the smooth-tasting wine enjoyed by pyramid-energy enthusiasts was aged by the pyramid in much the same manner.
- *Medical therapy.* One of the most frequent claims for pyramid energy is the therapeutic effect it has on people who subject themselves to its influence. Ultrasonics has claimed clinical successes in treating arthritis, muscular rheumatism, and sciatica.
- *The effect on bacteria and other microorganisms.* Ultrasonic radiation of sufficient intensity may destroy bacteria and other microorganisms; and if the intensity is low, growth is stimulated.

Karl Drbal, a Czechoslovak radio technician, began experimenting with pyramids in the late 1940s. The experiments disclosed some interesting phenomena that prompted Drbal to apply for a patent. With his colleagues, Drbal discovered that the sharpness of a razor blade was maintained longer than normal when the blade was kept inside a pyramid structure. In 1949, he submitted his "Pharaohs Shaving Device" to the patent office, but the officers did not take the application seriously and turned it down. With this rejection, Drbal staunchly resolved to determine how the pyramid shape worked, and then to explain it to the world in physical terms.

Collaborating with Drbal to find answers to the pyramid question were some of the finest metallurgical experts in Europe. Dr. Carl Benedicks of Stockholm, Sweden, experimented on the effect water had on steel. His results showed that water reduced the steel's hardness by as much as twenty-two percent. Benedicks' tests on worn razor blades revealed that moisture in the microcavities diminished the sharpness of the blade, and to stop this deterioration, it was necessary to dispel the water dipole molecules from the blade's edge. Benedicks' findings were that the pyramid shape created a resonance or vibratory field. The resonance caused dehydration of the water in the microcavities of the blade, thereby allowing the blade to retain its sharpness.[14]

Professors Born and Lertes of Germany demonstrated that the dipole molecules of water were affected by microwave energy inside a resonant cavity such as a pyramid and that microwaves of centimeter wavelength and their harmonics can generate an accelerated rotation of the dipole molecules, initiating the dehydration process.[15]

Assuming these reports are accurate, they are strong indications that the Great Pyramid conducts a broad range of vibrational frequencies through its mass. When I consider the mathematical comparison of the dimensions of the Great Pyramid with the dimensions of the Earth, I am led to conclude that this correspondence was no coincidence, but was in fact the expressed intention of the builders. If the dimensions of the Earth determine the wave characteristics of vibrations emanating from the core, then it would obviously be beneficial to incorporate these dimensions in a receiver of these vibrations. The receiver would respond harmonically to the influence of the vibrations and be in a state of resonance with them. The energy of the Earth is tremendous. The seismic disturbances around the globe (for instance, an estimated one million earthquakes occur annually) and the awesome power released by a volcanic eruption attest to the magnitude of this Earth energy. And these accumulated stresses are a constant factor in the Earth's evolution.

Energy is the basis of creating electricity that we can utilize, so how can we harness the power of an earthquake? Obviously, today, if that much energy were being drawn from the Earth through the Great Pyramid, tourists would not be parading through it every day. In order for the system to work, the pyramid would need to be mechanically coupled with the Earth and vibrating in sympathy with it. To do this, the system would need to be "primed"—we would need to initiate oscillation of the pyramid before we could tap into the Earth's oscillations. After the initial priming pulse, though, the pyramid would be coupled with the Earth and could draw off its energy. In effect, the Great Pyramid would feed into the Earth a little energy and receive an enormous amount out of it in return (see Figure 35).

How do we cause a mass of stone that weighs 5,273,834 tons to oscillate? It would seem an impossible task. Yet there was a man in recent history

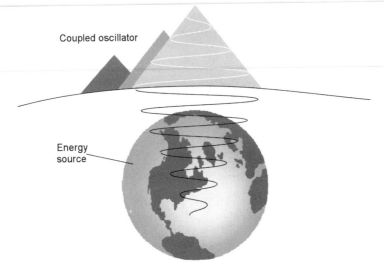

Coupled oscillator

Energy
source

A coupled oscillator will draw energy from the
source as long as the source continues to vibrate.

FIGURE 35. *Coupled Oscillator*

who claimed he could do just that! Nikola Tesla, a physicist and inventor
with more than six hundred patents to his credit—one of them being the
AC generator—created a device he called an "earthquake machine." By applying vibration at the resonant frequency of a building, he claimed he could
shake the building apart. In fact, it is reported that he had to turn his machine off before the building he was testing it in came down around him.
The New York *World-Telegram* reported Tesla's comments from a news brief-
ing at the hotel New Yorker on July 11, 1935:

> *I was experimenting with vibrations. I had one of my machines going
> and I wanted to see if I could get it in tune with the vibration of the
> building. I put it up notch after notch. There was a peculiar cracking
> sound.*
>
> *I asked my assistants where did the sound come from. They did
> not know. I put the machine up a few more notches. There was a
> louder cracking sound. I knew I was approaching the vibration of the
> steel building. I pushed the machine a little higher.*
>
> *Suddenly, all the heavy machinery in the place was flying around.*

I grabbed a hammer and broke the machine. The building would have been about our ears in another few minutes. Outside in the street there was pandemonium. The police and ambulances arrived. I told my assistants to say nothing. We told the police it must have been an earthquake. That's all they ever knew about it.

A reporter at that point asked Dr. Tesla what he would need to destroy the Empire State Building. Tesla replied:

Five pounds of air pressure. If I attached the proper oscillating machine on a girder that is all the force I would need, five pounds. Vibration will do anything. It would only be necessary to step up the vibration of the machine to fit the natural vibration of the building and the building would come crashing down. That's why soldiers break step crossing a bridge.[16]

Scientist Tom Bearden, in a paper to the International Tesla Society in 1988, went further with Tesla's research into using the Earth as a source for energy and proposed that "all that need be done to extract enormous energy is to input the 'grid signal' into the Earth, and receive the enormous 'plate signal' response. The standing S-wave is continuously replenished from the stress energy in the earth itself, so power may be extracted continuously."[17] Bearden cautioned, however, that his model was based on an "idealized isotropic medium, and our results eventually must be modified to take into account the earth's anisotropy."[18]

By applying Tesla's technology in the Great Pyramid, using alternating timed pulses at the apex of the pyramid and in the Subterranean Chamber—a feature, by the way, that all the Egyptian pyramids have—we may be able set into motion 5,273,834 tons of stone! If we have trouble getting the Great Pyramid going, there are three small pyramids nearby that we can start first to get things moving. Once the vibration of the pyramid is coupled to the vibration of the Earth, the transfer of energy from the Earth to the pyramid could be sustained until the process is reversed. Once the pyramid is coupled to the Earth, we would have to design a system that would do something with the energy. I propose that the Egyptians had in fact created such a system, using crystals and other natural elements. The secrets of ancient technology may be beginning to emerge in a fantastic and extraordinary way. Let me explain how.

Chapter Nine

THE MIGHTY CRYSTAL

nowing that we can design an object to respond sympathetically with the Earth's vibration, how do we utilize that energy? How can we turn it into usable electricity? We must, first of all, understand what a transducer is. Earlier on we discussed the piezoelectric effect vibration has on quartz crystal (refer to Figure 31 in chapter eight). Alternately compressing and releasing the quartz produces electricity. Microphones and other modern electronic devices work on this principle. Speak into a microphone and the sound of your voice (mechanical vibration) is converted into electrical impulses. The reverse happens with a speaker, where electrical impulses are converted into mechanical vibrations. As I mentioned, it also has been speculated that quartz-bearing rock creates the phenomenon known as ball lightning. It can do so because the quartz crystal serves as a transducer—it transforms one form of energy into another. When we understand the source of the energy and have the means to tap into it, all we need to do to convert the unlimited mechanical stresses therein into usable electricity is to utilize quartz crystals! As you might guess, the Great Pyramid contains quartz crystals, its own transducers.

Let me make no apology for the theory I am proposing. The Great Pyramid was a geomechanical power plant that responded sympathetically with the Earth's vibrations and converted that energy into electricity. They used the electricity to power their civilization, which included machine tools with which they shaped hard, igneous rock.

OK, you may say. Prove it! Just how does this power plant work? Well, let us start with the power crystals, or transducers. It so happens that the transducers for this power plant are an integral part of its construction and were designed to resonate in harmony with the pyramid itself and also with the Earth. The King's Chamber, in which a procession of visitors have noted unusual energetic effects and in which Tom Danley detected infrasonic vibrations (which I postulated were coming from the Earth) is, in itself, a mighty transducer.

In any machine there are devices that function to make the machine work. This machine was no different. Although the inner chambers and passages of the Great Pyramid seem to be devoid of what we would consider to be mechanical or electrical devices, there are devices still housed there that are similar in nature to mechanical devices created today. These devices also could be considered to be electrical devices in that they have the ability to convert or transduce mechanical energy into electrical energy. We might think of other examples, as the evidence becomes more apparent. The devices, which have resided inside the Great Pyramid since it was built, have not been recognized for what they truly were. Nevertheless, they were an integral part of this machine's function.

The granite out of which the King's Chamber is constructed is an igneous rock containing silicon-quartz crystals. This particular granite, which was brought from the Aswan quarries, contains fifty-five percent or more quartz crystal. Dee Jay Nelson and David H. Coville saw special significance in the builders' choice of granite for constructing the King's Chamber. They wrote:

This means that lining the King's Chamber, for instance, are literally hundreds of tons of microscopic quartz particles. The particles are hexagonal, by-pyramidal or rhombohedral in shape. Rhomboid crystals are six-sided prisms with quadrangle sides that present a parallelogram on any of the six facets. This guarantees that embedded within the granite rock is a high percentage of quartz fragments whose surfaces, by the law of natural averages, are parallel on the upper and lower sides. Additionally, any slight plasticity of the granite aggregate would allow a "piezotension" upon these parallel surfaces and cause an electromotive flow. The great mass of stone above the pyramid chambers presses downward by gravitational force upon the granite walls thereby converting them into perpetual electric generators.

The inner chambers of the Great Pyramid have been generating electrical energy since their construction 46 centuries ago. A man within the King's Chamber would thus come within a weak but definite induction field.[1]

While Nelson and Coville have made an interesting observation and speculation regarding the granite inside the pyramid, I am not sure that they are correct in stating that the pressure of thousands of tons of masonry would create an electromotive flow in the granite. The pressure on the quartz would need to be alternatively pressed and released in order for electricity to flow. The pressure they are describing would be static and, while it would undoubtedly squeeze the quartz to some degree, the electron flow would cease after the pressure came to rest. Quartz crystal does not create energy; it just converts one kind of energy into another. Needless to say, this point in itself leads to some interesting observations regarding the characteristics of the granite complex.

Above the King's Chamber are five rows of granite beams, making a total of forty-three beams weighing up to seventy tons each. The layers are separated by spaces large enough for the average person to crawl into. The red granite beams were cut square and parallel on three sides but were left seemingly untouched on the top surface, which is rough and uneven. Some of the beams even had holes gouged into their tops.

In cutting these giant monoliths, the builders evidently found it necessary to treat the beams destined for the uppermost chamber with the same respect as those intended for the ceiling directly above the King's Chamber. Each beam was cut flat and square on three sides, with the topside rough and seemingly untouched. Petrie wrote: "The roofing beams are not of 'polished granite,' as they have been described; on the contrary, they have rough-dressed surfaces, very fair and true so far as they go, but without any pretense to polish."[2] From his observations of the granite inside the King's Chamber, Petrie continued with those of the upper chambers: "All the chambers over the King's Chamber are floored with horizontal beams of granite, rough dressed on the under sides which form the ceilings, but wholly unwrought above."[3] These facts are interesting, considering that the beams directly above the King's Chamber would be the only ones visible to those entering the pyramid. Even so, the attention these granite ceiling beams received was nonetheless inferior to the attention commanded by the granite out of which the walls were constructed.

It is remarkable that the builders would exert the same amount of effort in finishing the thirty-four beams that would not be seen once the pyramid was built as they did the nine beams forming the ceiling of the King's Chamber, which would be seen. Even if these beams were imperative to the strength of the complex, deviations in accuracy would surely be allowed, making the cutting of the blocks less time consuming. Unless, of course, the builders were either using these upper beams for a specific purpose, or were using standardized machining methods that produced parts with little variation.

Traditional theory, proposed by Howard-Vyse and supported by Egyptologists, has it that the granite beams served to relieve pressure on the King's Chamber and allowed this chamber to be built with a flat ceiling. I disagree. The pyramid builders knew about and were already utilizing a design feature that was structurally sound on a lower level inside the pyramid. If we look at the cantilevered arched ceiling of the Queen's Chamber, we can see that it has more masonry piled on top of it than does the King's Chamber. The question could be asked, therefore, that if the builders had wanted to put a flat ceiling in this chamber, wouldn't they have needed to add only one layer of beams? For the distance between the walls, a single layer of beams in the Queen's Chamber, like the forty-three granite beams above the King's Chamber, would be supporting no more than their own weight (see Figure 36).

This leads me to ask, "Why does the King's Chamber need five layers of these beams?" From an architectural and engineering point of view, it is unnecessary to have so many monolithic blocks of granite in this structure. It is especially wasteful when we consider the amount of incredibly difficult work that must have been invested in quarrying, cutting, and transporting the stone from the Aswan quarries five hundred miles away—and then raising the beams to the 175-foot level of the pyramid. There is surely another reason for such an enormous effort and investment of time.

And look at the characteristics of these beams. Why cut them square and flat on three sides and leave them rough on the top? If no one is going to look at them, why not make them rough on all sides? Better still, why not make all sides flat? It would certainly make them easier to assemble. It is

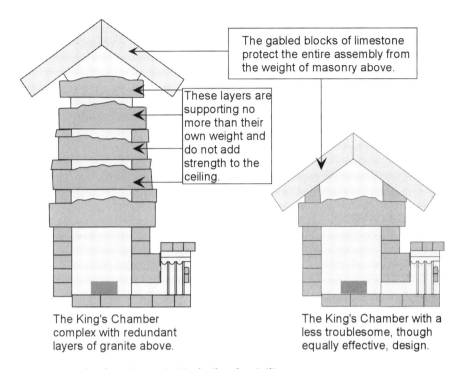

The gabled blocks of limestone protect the entire assembly from the weight of masonry above.

These layers are supporting no more than their own weight and do not add strength to the ceiling.

The King's Chamber complex with redundant layers of granite above.

The King's Chamber with a less troublesome, though equally effective, design.

FIGURE 36. *Redundant Granite in King's Chamber Ceiling*

clear, then, that the forty-three giant beams above the King's Chamber were not included in the structure to relieve this chamber from excessive pressure from above, but were included to fulfill a more advanced purpose. When we look at these beams with an engineer's eye, we can discern a simple, yet refined technology in this granite complex at the heart of the Great Pyramid, a technology that operated this power plant.

The giant granite beams above the King's Chamber could be considered to be forty-three individual bridges. Like the Tacoma Narrows Bridge, each one is capable of vibrating if a suitable type and amount of energy is introduced. If we were to concentrate on forcing just one of the beams to oscillate—with each of the other beams tuned to that frequency or its harmonic—the other beams would be forced to vibrate at the same frequency or a harmonic. If the energy contained within the forcing frequency was great enough, this transfer of energy from one beam to the next could affect the entire series of beams. A situation could exist, therefore, in which one individual beam, in the ceiling directly above the King's Chamber, could indirectly influence another beam in the uppermost chamber by forcing it to vibrate at the same frequency as the original forcing frequency or one of its harmonic frequencies. The amount of energy absorbed by these beams from the source would depend on the natural resonant frequency of the beam.

If this scenario is true, we have to consider the beams' ability to dissipate the energy they were subjected to, as well as their natural resonating frequency. If the forcing frequency (sound input) coincided with the beams' natural frequency (the beams were not restrained from vibrating), then the transfer of energy would be maximized. Consequently, so would the vibration of the beams.

We know that the giant granite beams above the King's Chamber have a length of seventeen feet (the width of the chamber), the entire length of which we assume can react to induced motion and vibrate without restraint. Some damping would occur if the beams' adjacent faces were so close that they rub together. However, if the beams vibrate in unison, it is possible that such damping would not happen. To perfect the ability of the forty-three granite beams to resonate with the forcing frequency, the natural frequency of each beam would have to be of the same frequency as the forcing frequency, or be in harmony with it.

It would be possible for us to tune a length of granite, such as those found in the Great Pyramid, by altering its physical dimensions. We could attain a precise frequency by either altering the length of the beam—as a guitarist alters the length of a guitar string—or by removing material from the beam's mass, as in the tuning of bells. (A bell is tuned to a fundamental hum and its harmonics by removing metal from critical areas.) If we would strike the beam, as one would strike a tuning fork, while it is being held in a position similar to that of the beams above the King's Chamber, we could induce oscillation of the beam. Then we could sample the frequency of the beam's vibration and remove more material until the correct frequency was reached.

Rather than suffering from a lack of attention, therefore, the rough top surfaces of those granite beams in the King's Chamber have been given more careful and deliberate attention and work than the beams' sides or bottoms. Before the ancient craftspeople placed them inside the Great Pyramid, each beam may have been "tested" or "tuned" by being suspended on each end in the same position that it would have once it was placed inside the pyramid (see Figure 37). The workers would then shape and gouge the topside of each beam in order to tune it before it was permanently positioned inside the pyramid. After cutting three sides square and true to each other, the remaining side could have been cut and shaped until it reached a specific resonating frequency. The removal of material on the upper side of the beam would take into consideration the elasticity of the beam, as a variation of elasticity might result in more material being removed at one point along the beam's length than at another. The fact that the beams above the King's Chamber are all shapes and sizes would support this speculation. In some of the granite beams, I would not be surprised if we found holes gouged out of the granite as the tuners worked on trouble spots. What we find in the King's Chamber, then, are thousands of tons of granite that were precisely tuned to resonate in harmony with the fundamental frequency of the Earth and the pyramid!

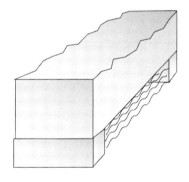

The tuning of a single beam can be accomplished by suspending it at the ends and selectively removing material from the top side until it "rings" at the correct frequency.

FIGURE 37. *Beam Tuning*

Smyth and Petrie unwittingly provided clues that this resonance theory not only may be plausible, but indeed may be probable. Both sought an explanation for the holes gouged near the ends of these granite beams. Smyth said, "These markings, moreover, have only been discovered in those dark holes or hollows, the so-called 'chambers,' but much rather 'hollows of construction,' broken into by Colonel Howard-Vyse above the 'King's Chamber' of the Great Pyramid. There, also, you see other traces of the steps of mere practical work, such as the 'bat-holes' in the stones, by which the heavy blocks were doubtless lifted to their places, and everything is left perfectly rough."[4] Rather than seeing them as holes used for lifting the blocks into place, Petrie speculated on an alternate reason for Smyth's so-called "bat-holes": "The flooring of the top chamber has large holes in it, evidently to hold the butt ends of beams which supported the sloping roof-blocks during the building."[5]

Neither Smyth's nor Petrie's explanations are particularly satisfactory. The most likely and logical reason for the holes gouged near the end of the beam may have been to strategically weaken the beam in order for it to respond more readily to sound input. According to Boris Said, who was with engineer Tom Danley when he conducted his acoustical tests inside the King's Chamber, the King's Chamber's granite beams resonated at a fundamental frequency and the entire structure of the chamber reinforced this frequency by producing dominant frequencies that created an F-sharp chord. Not surprisingly, the F-sharp chord is believed to be in harmony with the Earth. While testing for frequency, Danley placed accelerometers in the spaces above the King's Chamber, but I do not know whether he went as far as checking the frequency of each beam. Said said something in his interview with Art Bell that may be some indication of where Danley was heading with his research: He said that the beams above the King's Chamber were "like baffles in a speaker." Further research would need to be conducted before any assertion could be made as to the relationship these holes may have with tuning these beams to a specific frequency. However, when we consider the characteristics of the entire granite complex, along with other features found in the Great Pyramid, it seems clear that the results of this research will be along the lines of what I am theorizing.

Without confirmation that the granite beams were carefully tuned to respond to a precise frequency, I will infer that such a condition exists in light of what is found in the area. While I have not found any specific record of anyone striking the beams above the King's Chamber and measuring their resonant frequencies, there has been quite a lot written about the resonating qualities of the coffer inside the chamber itself. The coffer is said to resonate at 438 hertz and is at resonance with the resonant frequency of the chamber. This is easily tested and has been noted by numerous visitors to the Great Pyramid, including myself.

Another interesting discovery was made by the Schor expedition. This is a preliminary report, told to Art Bell by Boris Said: It was discovered that the floor of the King's Chamber does not sit on solid rock. Not only is the entire granite complex surrounded by massive limestone walls with a space between the granite and the limestone, the floor itself sits on what is characterized as "corrugated" rock (see Figure 38). Acoustical analysis of the floor of the King's Chamber (by Danley) revealed that the flooring blocks are not sitting on solid masonry. There are pockets beneath the floor that indicate that the support for the floor is "corrugated" like an egg carton, with the flooring sitting on nodes. In addition, the walls of the chamber do not sit on the granite floor, but are supported from the outside and sunk five inches below floor level. The entire complex is freestanding from the limestone masonry, has minimal damping of the floor, and is thus free to vibrate at peak efficiency. It is no wonder the entire chamber "rings" while tourists walk around inside!

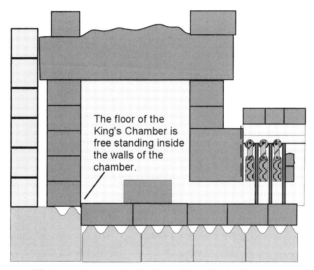

The floor of the King's Chamber is free standing inside the walls of the chamber.

What may lie beneath the floor of the King's Chamber

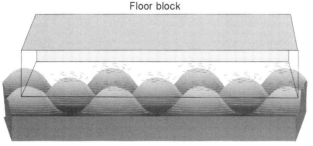

Floor block

Underlying support

FIGURE 38. *Floor of the King's Chamber*

159

The granite complex inside the Great Pyramid, therefore, is poised ready to convert vibrations from the Earth into electricity. What is lacking is a sufficient amount of energy to drive the beams and activate the piezoelectric properties within them. The ancients, though, had anticipated the need for more energy than what would be collected only within the King's Chamber. They had determined that they needed to tap into the vibrations of the Earth over a larger area inside the pyramid and deliver that energy to the power center—the King's Chamber—thereby substantially increasing the amplitude of the oscillations of the granite.

Modern concert halls are designed and built to interact with the instruments performing within. They are huge musical instruments in themselves. The Great Pyramid can be seen as a huge musical instrument with each element designed to enhance the performance of the other.

While modern research into architectural acoustics might focus predominantly upon minimizing the reverberation effects of sound in enclosed spaces, there is reason to believe that the ancient pyramid builders were attempting to achieve the opposite. The Grand Gallery, which is considered to be an architectural masterpiece, is an enclosed space in which resonators were installed in the slots along the ledge that runs the length of the gallery. As the Earth's vibration flowed through the Great Pyramid, the resonators converted the vibrational energy to airborne sound. By design, the angles and surfaces of the Grand Gallery walls and ceiling caused reflection of the sound, and its focus into the King's Chamber. Although the King's Chamber also was responding to the energy flowing through the pyramid, much of the energy would flow past it. The specific design and utility of the Grand Gallery was to transfer the energy flowing through a large area of the pyramid into the resonant King's Chamber. This sound was then focused into the granite resonating cavity at sufficient amplitude to drive the granite ceiling beams to oscillation. These beams, in turn, compelled the beams above them to resonate in harmonic sympathy. Thus, with the input of sound and the maximization of resonance, the entire granite complex, in effect, became a vibrating mass of energy.

Sound farfetched? Not if we realize that many isolated aspects of this proposed phenomenon have been noted by visitors and researchers over the years. In particular, the acoustic qualities of the design of the upper chambers of the Great Pyramid have been referenced and confirmed by numerous visitors since the time of Napoleon, whose men discharged their pistols at the top of the Grand Gallery and noted that the explosion reverberated into the distance like rolling thunder. Strike the coffer inside the King's Chamber and you will hear a deep bell-like sound of incredible and eerie beauty. It has been a practice over the years for the Arab guides to demonstrate this resonating sound to the tourists they guide through the Great Pyramid. This sound was even included on Paul Horn's album, *Inside the Great Pyramid*. After being advised of the significant pitch produced by striking the coffer, and the chamber's response to this pitch, Horn took along a device that would allow him to replicate the exact pitch and frequency. Horn struck the coffer and tuned his flute to the tone that was emitted, which turned out to be the note A—which vibrates at 438 cycles per second. In a fascinating booklet about his experiment at the Great Pyramid, Horn described his experience in the inner chambers: "The moment had arrived. It was time to play my flute. I thought of Ben Peitcsh from Santa Rosa, California, and his suggestions to strike the coffer. I leaned over and hit the inside with the fleshy part of the side of my fist. A beautiful round tone was immediately produced. What a resonance! I remember him also saying when you hear that tone you will be 'poised in history that is ever present.' I took the electronic tuning device I had brought along in one hand and struck the coffer again with the other and there it was—'A' 438, just as Ben predicted. I tuned up to this pitch and was ready to begin. [The album opens with these events so that you can hear all of these things for yourselves.]"[6]

And, indeed, the sound that Paul Horn brought into my living room was most fascinating. Listening to it, I could understand why so many people develop feelings of reverence when hearing it, for it has a most soothing effect on the nerves. For this alone, the album was worth the price. Horn himself described the effects this sound had on him: "Sitting on the floor in front of the coffer with the stereo mike in the centre of the room, I began to play, choosing the alto flute to begin with. The echo was wonderful, about eight seconds. The chamber responded to every note equally. I waited for the echo to decay and then played again. Groups of notes would suspend and all come back as a chord. Sometimes certain notes would stick out more than others. It was always changing. I just listened and responded as if I were playing with another musician. I hadn't prepared anything specific to play. I was just opening myself to the moment and improvising. All of the music that evening was this way—totally improvised. Therefore, it is a true expression of the feelings that transpired."[7]

After noting the eerie qualities of the King's and Queen's Chambers, Horn went out onto the Great Step at the top of the Grand Gallery to continue his sound test. The Grand Gallery, he reported, sounded rather flat compared with the other chambers, but then he heard something remarkable: The music he was playing was coming back to him clearly and distinctly from the King's Chamber. The sound was going out into the Grand Gallery and was being reflected through the passageway and reverberating inside the King's Chamber!

Horn does not attempt to explain this acoustical phenomenon, but it is tied in with the phenomena noted inside the King's Chamber. It would follow, therefore, that the coffer inside the King's Chamber was specifically tuned to a precise frequency, and that the chamber itself was scientifically engineered to resonate in harmony with sound waves that were generated in the Grand Gallery and focused into it. Perhaps these observations will provide an answer to Horn's experience and to a mystery that Petrie puzzled over at great length. He discovered a flint pebble under the coffer, after he raised it, and instead of dismissing it as debris or otherwise, he mulled over its significance. He wrote:

The flint pebble that had been put under the coffer is important. If any person wished at present to prop the coffer up, there are multitudes of stone chips in the pyramid ready to hand. Therefore, fetching a pebble from the outside seems to show that the coffer was first lifted at a time when no breakages had been made in the pyramid, and there were no chips lying about. This suggests that there was some means of access to the upper chambers, which are always available by removing loose blocks without any forcing. If the stones at the top of the shaft leading from the subterranean part to the gallery had been cemented in place, they must have been smashed to break through them, or if there were granite portcullises in the Antechamber, they must also have been destroyed; and it is not likely that any person would take the trouble to fetch a large flint pebble into the innermost part of the Pyramid, if there were stone chips lying in his path.[8]

If Petrie says that something is important, I tend to take notice of what he is talking about. Nonetheless, I am not convinced that this pebble could *not* have been brought into the King's Chamber long after the Great Pyramid was built and used to prop up the coffer so that it could be moved. On the other hand, Petrie does pose another alternative that deserves some speculation, and I cannot help wondering if it is possible that the pebble served a greater purpose for those who placed it there. If we had just manufactured an object like the coffer and had it tuned to vibrate at a precise frequency, we would know that to set it flat on the floor would dampen the vibrations somewhat. By raising one end of the coffer onto the pebble, however, it could vibrate at peak efficiency.

Another unique feature of the Grand Gallery, which needs to be confirmed by on-site inspection, is the approximate angle that is achieved by it having a ratchet-style ceiling. The problem with coming up with an accurate calculation of the true angle of the overlapping stones in this ceiling is that there is conflicting data from the only two researchers I have found who pay these overlaps any close attention. However, preliminary calculations are interesting to say the least. According to Smyth, the angle of the Grand Gallery is 26°17'37", the height of the Grand Gallery is between 333.9 inches and 346.0 inches, and he counted thirty-six overlaps in the 1,844.5-inch length of the ceiling.[9] Writing about his observations and measurements of the tilting tiles, Smyth said, ". . . when I was actually pushing up the point of a long measuring rod, against the roof stones, differences were found to so great an extent as 12.1 inches, which I did attribute chiefly to that very cause [that the tiles did indeed tilt]."[10] With the overlaps estimated to have approximately a twelve-inch tilt, the surface of the overlapping stones in the ceiling yields an amazing approximation to a 45-degree angle from a vertical plane (135 degrees polar coordinates, given that the ends of the gallery are 90 degrees). With this tilt of the roof tiles, a sound wave traveling vertically to the roof would be reflected off the tiles at a 90-degree angle and would travel in the direction of the King's Chamber (see Figure 39).

This gives pertinence to another report, which did not receive much

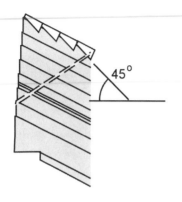

The ceiling tiles in the Grand Gallery tilt, resulting in an angle of approximately 45 degrees.

FIGURE 39. *Roof Tiles of the Grand Gallery*

attention at the time it was made. It has been reported that Al Mamun's men had to break a false floor out of the Grand Gallery, and as they broke one stone out, another slid down in its place. This is a sketchy bit of information that would require further research, if further research is possible, because Al Mamun's men were tearing out so much limestone that little attention was given to this incident. However, what if that stone slid because the tiled floor in this gallery had a rachet style that matched the style of its ceiling? Much of the stone that Al Mamun cut out of the Ascending Passage was dropped down the Descending Passage. Later explorers, such as Caviglia, Davison, and Petrie, eventually cleared this passage of all debris, and most of this debris was dumped on the traditional rubbish pit on the north and east sides of the Great Pyramid. Petrie reported finding inside the Great Pyramid a prism-shaped stone that had a half-round groove running its length. In the Descending Passage he found a block of granite that was 20.6-inches thick with a tube-drilled hole cut through the thickness on one edge. Where this granite came from and what purpose it was used for in the Great Pyramid was a mystery to Petrie, who wrote, "What part of the Pyramid this can have come from is a puzzle; nothing like it, and no place for it, is known."[11] With more significant findings attracting researchers' attention, though, it is not surprising these details were not given much consideration.

It has been assumed that the rachet-style ceiling in the Grand Gallery was so designed to prevent an accumulation of forces bearing down the angle of the gallery and pressing on the lower pieces. Yet other angled passages in the Great Pyramid, such as the Ascending and Descending Passages, have flat ceilings, so I am left to conclude that this feature was, indeed, specifically designed for an acoustical purpose.

Even if some of the masonry clues are now lost to us, it may be possible for an acoustical engineer to confirm that the Grand Gallery indeed reflected sound in the manner proposed by examining only its dimensions and angles. Perhaps this book will encourage an engineer to create a computer model of the Grand Gallery and perform an analysis by simulating the movement of sound within the cavity. Though I have attempted to find some means to accomplish this, I have not been able to find anybody with access to a supercomputer who is willing to do the work, and the software needed to perform the analysis has not, to my knowledge, been published for a micro-computer yet.

We can also extrapolate other information about acoustical devices that are obviously no longer in place within the Grand Gallery. Knowing that the King's Chamber will respond to sound of a specific frequency, thereby trans-ducing that energy into electrical energy, I theorized earlier that the Grand Gallery housed resonators that converted the coupled Earth/pyramid vibra-tion to airborne sound. The existence of resonators in this gallery is pre-dicted by what has been found inside the King's Chamber and the design of, and phenomena noted in, the Grand Gallery. The mystery of the twenty-seven pairs of slots in the side ramps is logically explained if we theorize that each pair of slots contained a resonator assembly and the slots served to lock these assemblies into place. The original design of the resonators will always be open to question; however, if their function was to efficiently respond to the Earth's vibration, then we can surmise that they might be similar to a device we know of today that has a similar function—a Helmholtz resonator.

A Helmholtz resonator responds to vibrations and actually maximizes the transfer of energy from the source of the vibrations. The resonator is normally made out of metal, but it can be made out of other materials. A classic example of a Helmholtz resonator is a hollow sphere with a round opening that is 1/10 to 1/5 the diameter of the sphere. The size of the sphere determines the frequency at which it will resonate. If the resonant frequency of the resonator is in harmony with a vibrating source, such as a tuning fork, it will draw energy from the fork and resonate at greater amplitude than the fork is able to without its presence. It forces the fork to greater energy output than what is normal, or "loads" the fork. Unless the energy in the fork is replenished, its energy will be exhausted quicker than it nor-mally would be without the Helmholtz resonator. But as long as the source

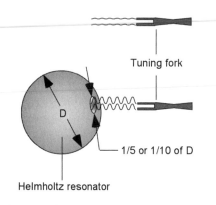

Tuning fork

D

1/5 or 1/10 of D

Helmholtz resonator

By virtue of its design, the Helmholtz resonator, over time, draws more energy from a vibrating source, such as a tuning fork, than what the source will give up naturally.

FIGURE 40. *Helmholtz Resonator*

continues to vibrate, the resonator will continue to draw energy from it at a greater rate (see Figure 40).[12]

To extrapolate further we could say that each resonator assembly that was installed in the Grand Gallery was equipped with several Helmholtz-type resonators that were tuned to different harmonic frequencies. In a series of harmonic steps, each resonator in the series responded at a higher frequency than the previous one. In a manner similar to the King's Chamber's response to energy inputs—its creation of an F-sharp chord—these resonators raised the frequency of the vibrations coming from the Earth. To increase the resonators' frequency, the ancient scientists would have made the dimensions smaller, and correspondingly reduced the distance between the two walls adjacent to each resonator. In fact, the walls of the Grand Gallery actually step inward seven times in their height and most probably the resonators' supports reached almost to the ceiling. At their base, the resonators were anchored in the ramp slots. Not surprisingly, there is additional evidence in the Grand Gallery to support this premise, especially in a design feature of the gallery that is seldom given much thought. This is a groove, or slot, cut along the length of the second layer of the corbelled wall. This groove suggests the resonators were held in place inside the Grand Gallery and positioned, or keyed, into the structure by first being installed into the ramp slots and then held in a vertical position with "shot" pins in the groove. Once the resonator assemblies were positioned and locked into place, the angle of the slot effectively prevented them from moving (see Figure 41). The vertical supports for the Egyptian resonators were most likely made of wood because it is one of the most efficient responders to vibration; and, as we will discuss in chapter twelve, their disappearance from the pyramid can be easily explained. In crafting the resonators out of wood, the ancient Egyptians made a natural and logical choice, as the wood probably emitted a humming sound itself.

Prior to my visit to Egypt in 1986, I had speculated that the slots along the Grand Gallery floor anchored wooden resonators, but that these devices were balanced in a vertical orientation reaching almost to the full height of the gallery. If this speculation were true, it would logically follow that the geometry of the twenty-seven pairs of slots would provide proof. The bottom of the slots might have been parallel to the horizontal plane rather than

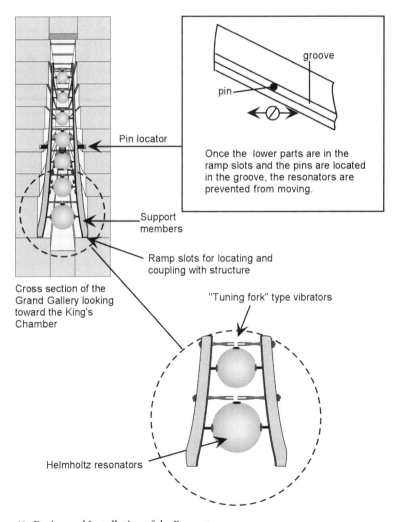

FIGURE 41. *Design and Installation of the Resonators*

parallel with the angle of the gallery, and the side walls of the slots might have been vertical to a horizontal plane rather than perpendicular to the angle of the gallery. This was a significant detail and a simple one to check out.

My first trek inside the Great Pyramid in 1986 did not reveal anything about the geometry of these slots because they were filled with dirt and debris. The following day I set out with a soup spoon that I had "borrowed" from the hotel restaurant. Digging out the dirt and debris, with tourists and guides looking at me like I was crazy (actually, it was probably illegal to do this as you need special permission to carry out excavations in Egypt), I finally came to the bottom of one slot and it was as I predicted it would be: parallel to the horizontal, and the sides of the slot perpendicular to the horizontal. Other slots were perpendicular to the horizontal as well, though some of them had bottoms that were parallel with the gallery floor. In either scenario, it appears that the slots were prepared to accommodate a vertical structure, rather than to restrain weight that would exert shear pressure from the side (see Figure 42).

In attempting to determine the design and materials the resonators were made of, we must consider evidence from artifacts that are not inside the

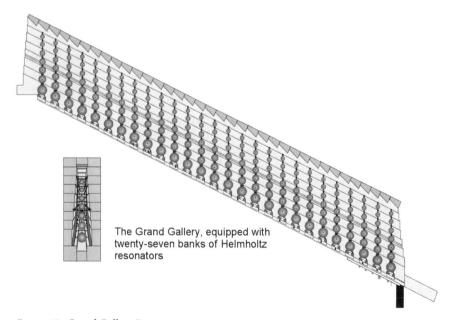

The Grand Gallery, equipped with twenty-seven banks of Helmholtz resonators

FIGURE 42. *Grand Gallery Resonators*

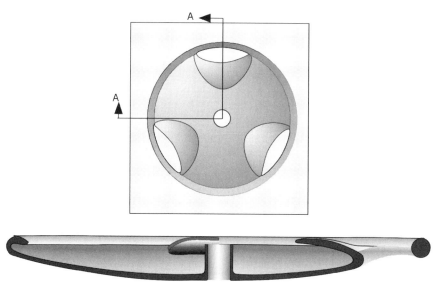

Section A-A
The approximate characteristics of
a schist bowl in the Cairo Museum

Photograph Courtesy of Robert McKenty

FIGURE 43. *Schist Bowl*

pyramid but are housed within the Cairo Museum, where the most remarkable feats of machining in the world can be found. In the museum's collection are stone jars and bowls so finely machined and perfectly balanced that
they inspire awe and wonder. One bowl in particular, a schist bowl with three
lobes folded toward the center hub, is an incredible piece of work. If the artisans used ultrasonics and sophisticated machinery, I can understand how it

could have been made; but even if they used those methods, the purpose for creating such a complex piece, for what is assumed to have been a domestic purpose, has long escaped me (see Figure 43). Other vases with small necks that open into a wide round belly on the inside have confounded other researchers, such as Graham Hancock, who wrote, "During my travels in Egypt I had examined many stone vessels—dating back in some cases to pre-dynastic times—that had been mysteriously hollowed out of a range of materials such as diorite, basalt, quartz crystal and metamorphic schist. For example, more than 30,000 such vessels had been found in the chambers beneath the Third Dynasty Step Pyramid of Zoser at Saqqara. That meant that they were at least as old as Zoser himself (i.e. around 2650 BC)."[13]

Regardless of the age of these artifacts, the technical accomplishment of the artisans who created the vases and bowls does not support the notion that they worked with primitive tools. I am in full agreement with Hancock, who pondered on an unimaginable technology and said, "Why unimaginable? Because many of the vessels were tall vases with long, thin, elegant necks and widely flared interiors, often incorporating fully hollowed-out shoulders. No instrument yet invented was capable of carving vases into shapes like these, because such an instrument would have had to have been narrow enough to have passed through the necks and strong enough (and of the right shape) to have scoured out the shoulders and the rounded interiors. And how could sufficient upward and outward pressure have been generated and applied within the vases to achieve these effects?"[14]

The questions Hancock posed are legitimate. As a machinist, I have created smaller products with similar geometry, for the aerospace industry. The technique I used to hollow out these modern stainless steel artifacts on a lathe involved drilling a hole and then using a series of L-shaped tools. After the first L-shaped tool had been used to open up the "chamber" bore, I held each successive tool in my hand and "snaked" the L through the neck into the piece before clamping the tool in the tool-holder. Once it was firmly clamped, I had to make sure that it was positioned correctly within the piece before the lathe began to rotate. It was a tricky operation that required the use of a small mirror and a powerful light to see inside the workpiece (see Figure 44).

The work I performed in steel was easy compared to the granite, dior-

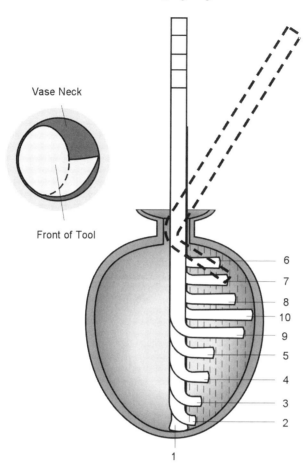

Vase Neck

Front of Tool

Tool numbers 1 through 10 are used in succession to hollow out the inside of a vase, such as the one illustrated. The dotted lines show the amount of material removed by one tool before another tool is needed. Ultrasonic assisted machining may have been employed in removing the material, as it would be difficult, if not impossible, for the tool to transfer adequate conventional machining forces from the tool holder to the cutting edge. Removing tool marks from the inside of the vase could be accomplished by using a slurry of small abrasive rock and water--similar to what lapidaries use today--and by rotating the bowl on its outside surface.

FIGURE 44. *Machining a Swan-neck Vase*

ite, basalt, and metamorphic schist that were cut by the ancient Egyptians. I was extremely puzzled and could not undertand why they went to such trouble to create vases in this manner. We certainly do not go to this trouble to create vases today. Their purpose was obviously very important to have

Photograph courtesy of Robert McKenty

Is this fine stonework in the Cairo Museum actually a Helmholtz resonator?

FIGURE 45. *Vessel with Horn Attached*

Photographs courtesy of Robert McKenty

FIGURE 46. *Precision Machined and Balanced Bowl*

required such an effort. It then occurred to me that perhaps these stone artifacts were not domestic vases at all, but were used in some other way. Perhaps they were being used to convert vibration into airborne sound. Given their shape and dimensions—and the fact that there were 30,000 of them found in chambers underneath the Step Pyramid—are these vessels the Helmholtz resonators we are looking for? As if to provide us with clues, one of the bowls in the Cairo Museum has a horn attached to it, and one of the bowls does not have handles normally seen on a domestic vase, or urn, but has trunnion-like appendages machined on each side of it. These trunnions would be needed to hold the bowl securely in a resonator assembly (see Figures 45 and 46).

Heading back to the Great Pyramid, we find more evidence for our theory. The enigmatic Antechamber has been the subject of much consternation and discussion. Ludwig Borchardt, director of the German Institute in Cairo (circa 1925), proposed that a series of stone slabs were slid into place after Khufu had been entombed. He theorized that the half-round grooves in the granite wainscoting supported wooden beams that served as windlasses to lower the blocks (see Figure 47).

Borchardt may not have been far off with his analysis of the mechanism that was contained within the Antechamber. After building the resonators and installing them inside the Grand Gallery, the ancient Egyptians would have wanted to focus a sound of specific frequency, that is, a pure or harmonic chord, into the King's Chamber. They would be assured of doing so if they installed an acoustic filter between the Grand Gallery and the King's Chamber. By installing baffles inside the Antechamber, sound waves traveling from the Grand Gallery through the passageway into the King's Chamber would be filtered as they passed through, allowing only a single frequency or harmonic of that frequency to enter the resonant chamber. Those sound waves with a wavelength that did not coincide with the dimensions between

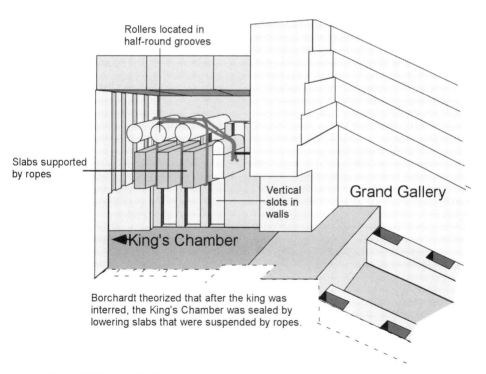

Rollers located in
half-round grooves

Slabs supported
by ropes

Vertical
slots in
walls

Grand Gallery

King's Chamber

Borchardt theorized that after the king was
interred, the King's Chamber was sealed by
lowering slabs that were suspended by ropes.

FIGURE 47. *Borchardt's Theory*

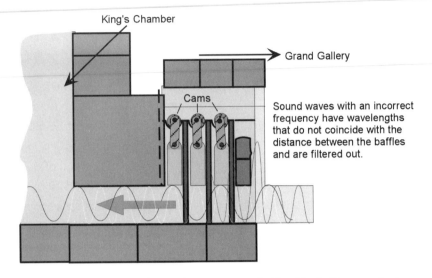

King's Chamber

Grand Gallery

Cams

Sound waves with an incorrect frequency have wavelengths that do not coincide with the distance between the baffles and are filtered out.

The Antechamber serving as an acoustic filter. The baffles are raised or lowered to "tweak" the system and to maximize its throughput.

FIGURE 48. *Acoustic Filter*

the baffles would be filtered out, thereby ensuring that no interference sound waves could enter the resonant King's Chamber to reduce the output of the system (see Figure 48).

To explain the half-round grooves on the west side of the Antechamber and the flat surface on the east, we could speculate that when the installation of these baffles took place, they received a final tuning or "tweaking." This may have been accomplished by using cams. By rotating the cams, the off-centered shaft would raise or lower the baffles until the throughput of sound was maximized. A slight movement may have been all that was necessary. Maximum throughput would be accomplished when the ceiling of the first part of the passageway (from the Grand Gallery), the ceiling of the passage-way leading from the acoustic filter to the resonant King's Chamber, and the bottom surface of each baffle were in alignment. The shaft suspending the baffles would have then been locked into place in a pillar block located on the flat surface of the wainscoting on the opposite wall.

During my conversations with Stephen Mehler and acoustical engineer Robert Vawter, I discussed my theory of the Antechamber. Vawter confirmed my analysis that it was used as an acoustic filter and agreed that further studies are needed to quantify the exact physics employed, via "back engineering" the dimensions of the King's Chamber complex.

So, what if my theory is correct and the Great Pyramid was a resonance chamber power plant? The next questions would be how the Egyptians controlled the vibrations to avoid an overload or destructive wave like that which downed the Tacoma Narrows Bridge. Because a vibrating system can eventually destroy itself if there is no means to draw off or dampen the energy, the ancient Egyptians would have had to find some way to control the level of energy at which the system operated. Because the output of the resonant cavity would draw off the energy only up to a certain level—that being the maximum amount the granite complex could process—there would have to be some means of controlling the energy as it built up inside the Grand Gallery.

Normally there are three ways to prevent a vibrating system from running out of control. First, you can shut off the source of vibration. Obviously if the source of the vibration in the Great Pyramid was the Earth itself, the ancient Egyptians could not do this. Second, you can reverse the process that couples the resonator to its source. Third, you can contrive a means to keep the vibrations at a safe level. Because the Great Pyramid's source of vibration was the Earth, options two and three are the obvious choices to solve the problem. We will address number three because, in order for the power plant to continue to function, there must be a constant supply of energy. There are two ways to control a constant vibration. One is to dampen it and the other is to counteract it with an interference wave that cancels out the vibration. Physically dampening the vibration would be impractical, considering the function of the Great Pyramid as a machine. The dampening would not always be necessary, unlike the dampening needs of a bridge, and indeed would have an adverse effect on the efficiency of the machine. It most probably would also involve building moving parts, like dampeners in a piano. Faced with these considerations, I looked more closely at the possibility that the ancient Egyptians may have cancelled out excess vibrations by using an out-of-phase interference sound wave. My inquiry began with the Ascending Passage, for it is the only feature inside the Great Pyramid that contains "devices" that are directly accessible from the outside. If the opera-

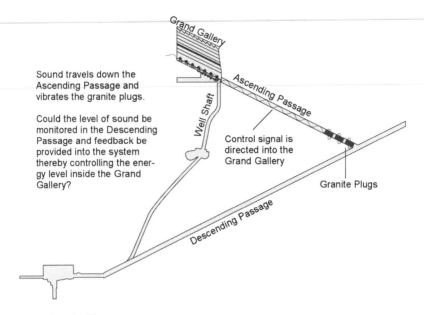

Sound travels down the
Ascending Passage and
vibrates the granite plugs.

Could the level of sound be
monitored in the Descending
Passage and feedback be
provided into the system
thereby controlling the energy level inside the Grand
Gallery?

Control signal is
directed into the
Grand Gallery

Granite Plugs

FIGURE 49. *Granite Plugs*

tors of the machine found it necessary to cancel out vibrations, they would
have to be able to quickly respond to this need from the machine's exterior.
I call the granite plugs inside this passage "devices" for the same reason I call
the granite beams above the King's Chamber devices—for they are over-
designed and over-crafted for their supposed use (see Figure 49). In the case
of the Ascending Passage, it was not necessary to use granite to block this
passage if the ancient Egyptians wanted to keep tomb robbers out—limestone would have been sufficient. Realistically, limestone would have been
better. The very fact that the stones "blocking" this passage are granite contradicts their effectiveness at securing the inner chambers from robbers, which
is the orthodox interpretation for their use. In fact, they had the opposite
effect, drawing Al Mamun's attention to the existence of the Ascending Passage and, subsequently, to the entire internal arrangement of passages and
chambers. No, these granite plugs had to have another reason for being there!

As I studied them, and envisioned their use, I realized that they may
have been built into the structure to perform two critical roles. One would
be to provide feedback to the power plant operators by responding to the
sound that was generated inside the Grand Gallery and traveling down the

Ascending Passage. The operators of this machine could have attached vibration sensors—much like the accelerometers that Danley used—to the bottom granite plug that is accessible from the Descending Passage, and thereby had the ability to monitor the energy level inside the Grand Gallery. The other task these granite plugs may have performed could have been to respond to vibration from equipment in the Descending Passage and transmit out-of-phase interference sound waves into the Ascending Passage and Grand Gallery to prevent vibrations within from reaching destructive levels. These functions explain not only the builders' reasons for selecting granite instead of limestone for these blocks but also the means by which the ancient Egyptians controlled the energy level of the system. We also could speculate that by directing a signal of the correct frequency up the Ascending Passage, the Egyptians may have been able to prime the system. In other words, the entire system would be forced to vibrate, and once in motion, it would draw energy from the Earth with no further input.

When Petrie examined these blocks, he noted that the adjoining faces of the blocks were not flat but had a wavy finish ± .3 inches. He wrote:

> These plug-blocks are cut out of boulder stones of red granite, and have not the faces cut sufficiently to remove the rounded outer surfaces at the corners: also the faces next to each other are never very flat, being wavy about plus or minus .300 inch. These particulars I was able to see, by putting my head in between the rounded edges of the 2nd and 3rd blocks from the top, which are not in contact; the 2nd having jammed tight 4 inches above the 3rd. The present top one is not the original end; it is roughly broken, and there is a bit of granite still cemented to the floor some way farther south of it. From appearances there I estimate that originally the plug was 24 inches beyond its present end.[15]

I was unable to confirm this when I was in Egypt because the blocks exposed by Al Mamun's tunnel had slipped since Petrie's day and are now resting against each other. (An interesting aside about Petrie's observations is that they fully dismiss and put to rest orthodox theories regarding the granite plugs being slid down the Ascending Passage. Some granite, in Petrie's words, was "still cemented to the floor. . . ." This is indisputable proof that the granite plugs were positioned as the Great Pyramid was being built.) However, Petrie's observation makes for interesting speculation within the context of my theory. Were the faces of the blocks cut specifically to modify sound waves? Could the Ascending Passage serve to direct an interference out-of-phase sound wave into the Grand Gallery, thereby controlling the level of energy in the system? As I sought to answer these questions, more mysteries about the Great Pyramid revealed themselves—and began to yield to the possible answers provided by my theory.

Chapter Ten

AN AMAZING MASER

he elements of this power system that we have discussed so far are the crystal transducers contained within the King's Chamber granite, the chamber's tuned harmonic characteristics with the Earth, the acoustical properties of the Grand Gallery, and the design features (acoustic filters) of the Antechamber. How each of these elements fit together, with each reinforcing the other, becomes more significant as other elements of this energy system are brought to bear.

Obviously the above features are not all that are needed to make this system work. After transducing mechanical energy into electrical energy, there has to be a medium through which the electricity can flow and be utilized. In a modern power plant, steam passes across turbine blades causing rotation of a generator that stimulates electron flow through copper wires. In *this* power plant the vibrations from the Earth cause oscillations of the granite within the King's Chamber, and this vibrating mass of igneous, quartz-bearing rock influences the gaseous medium contained within the chamber. Currently this gaseous medium is air, but when this power plant operated, it was most likely hydrogen gas that filled the inner chambers of the Great Pyramid. The Queen's Chamber holds evidence that it was used to produce hydrogen, and we will look at this evidence in the next chapter. But first, we must look at the evidence for the technology that utilized this gas.

To maximize the output of the system, the atoms comprising this gaseous medium contained within the chamber should have a unique characteristic—the gas's natural frequency should resonate in harmony with the entire system. To be more accurate, the resonance of the chamber, which can be adjusted, should resonate in harmony with the frequency of hydrogen, which does not change. Under these conditions, the hydrogen atoms would more efficiently absorb the energy generated within the chamber. Atomic hydrogen is the simplest atom, with one proton and one electron, and is responsible for the emission of microwave energy in the universe. This microwave background radiation is left over from the Big Bang and was first observed in 1965 by Arno Penzias and Robert Wilson at the Bell Telephone Laboratories in Murray Hill, New Jersey. The radiation has virtually the same energy or temperature in all directions in the sky. If we look up at the night sky, we can see many bright regions with clusters of stars and bright planets piercing the black backdrop of the universe. The cosmic microwave background temperature, however, varies very little across the sky. This signal from the Big Bang is constant and has been bombarding the Earth since it first began to form.[1] It is this signal from the universe that is an integral element in making the Giza power plant work. In order to understand how this could be, we need to understand how a maser works.

MASER is an acronym for Microwave Amplification through Stimulated Emission Radiation. The maser was developed before the laser—another acronym, meaning Light Amplification through Stimulated Emission Radiation—which, when it was first developed, was called an optical maser. So let us take a brief look at how lasers and masers operate.

To understand how a maser works, it would be best to describe something we can all see—light. Most of the light we are familiar with is incoherent light. The light that fills our homes from a fluorescent tube after the flip of a switch is the result of an electrical discharge energizing the atoms in a gaseous medium and pumping the electrons into a higher orbit around the proton. The electrons cannot be sustained at this higher energy level indefinitely, and will eventually fall back to their original position, or "ground state," releasing a packet of electromagnetic energy in the process. This packet of energy is called a photon. These photons are what we see as light, and the properties of the photons, with respect to wavelength and frequency, depend on the atoms in the gas. We recognize these properties by the color of the light. With fluorescent, the emission of photons is random and the pho-

Incoherent light–photons traveling in all directions

FIGURE 50. *The Fluorescent Light*

tons travel in unpredictable directions (see Figure 50). The number of photons that are constantly being emitted is large enough that they will travel every conceivable direction possible, thereby illuminating a room.

The uniqueness of the laser is based on the assumption that while photons in a light tube will travel in every direction, one of those directions will be along the length of the tube, parallel to its axis. If we place mirrors at both ends of the tube, therefore, and align these mirrors perfectly parallel with each other, the photons will reflect off the mirrors, back along the same axis they are traveling.

This is where the "stimulated emission" part of the laser comes in to play. A photon traveling back along its axis will encounter an atom with an electron in a higher energy state. The photon collides with the electron and forces it down to a lower energy state, which stimulates the release of another photon. Now there are two photons traveling together along the axis of the tube to the mirror at the other end. The train of two photons then becomes four photons, then eight, and so on (see Figure 51). With photons traveling at 186,282 miles per second, across a relatively short distance, the built up laser energy is almost instantaneous.[2]

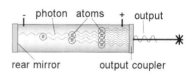

A photon collides with an energized atom and stimulates the emission of another photon. Each photon travels in phase "lock step" with the other photons.

FIGURE 51. *Principles of a LASER*

A laser, however, is subject to the same runaway vibrations that we discussed earlier, as in the Tacoma Narrows Bridge disaster. The laser cavity is a resonator, and there has to be some release of the energy or the resonator will be destroyed. To accomplish this release, one mirror is coated with a material that allows a percentage of the laser energy to be transmitted through the substrate, while the remainder is fed back into the cavity to sustain the "lasing" action.

The light beam that is emitted from a laser is coherent, collimated (does not spread like the light from a flashlight), and monochromatic. In other words, light of single frequency, or color, passes through the end mirror in a tight pencil beam. Light travels in waves, and the waves of the laser beam are of the same wavelength, and are in phase, or "lock stepped" with each other. This is the "amplification" part of the laser. Because the beam is coherent

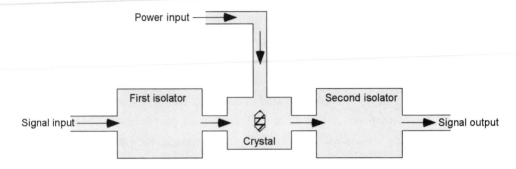

A Three-stage microwave amplifier

Source: Van Nostrand's Scientific
Encyclopedia 5th ed.

FIGURE 52. *Microwave Amplifier*

and collimated, the light cannot be seen until it strikes an object, such as a presentation screen, or as smoke in the air.

We know that a lightbulb is incoherent light energy, and, similarly, we can point to a microwave oven as an example of incoherent microwave energy. The maser operates a little differently from the laser in that its photons are in a different part of the electromagnetic spectrum—nevertheless, its principle of operation is similar. There are many different designs and iterations of designs for both lasers and masers (see Figure 52). In the Great Pyramid, there is evidence that strongly indicates that the ancient Egyptian engineers and designers knew about and utilized the principles of a maser to collect the energy that was being drawn through the pyramid from the Earth and deliver it to the outside. This evidence can be found in the King's Chamber.

A power plant could be described as a large engine. As with any other kind of engine, fuel is fed into it and is converted into energy. This energy can then be converted to other forms of energy—such as mechanical or electrical energy—that is then drawn off and used in whatever way is suitable. The key to converting or transducing hydrogen gas to usable power in the Giza power system was the introduction of acoustical vibration of the correct frequency and amplitude. (The amplitude is the amount of energy contained within a sound wave.) Based on the previous evidence, sound must have been focused into the King's Chamber to force oscillations of the granite, creating in effect a vibrating mass of thousands of tons of granite. The frequencies inside this chamber, then, would rise above the low frequency of the Earth—through a scale of harmonic steps—to a level that would excite the hydrogen gas to higher energy levels. The King's Chamber is a technical wonder. It is where Earth's mechanical energy was converted, or transduced, into usable power. It is a resonant cavity in which sound was focused. Sound roaring through the passageway at the resonant frequency of this chamber—or its harmonic—at sufficient amplitude would drive these granite beams to vibrate in resonance. Sound waves not of the correct frequency would be filtered in the acoustic filter, more commonly known as the Antechamber (see Figure 53).

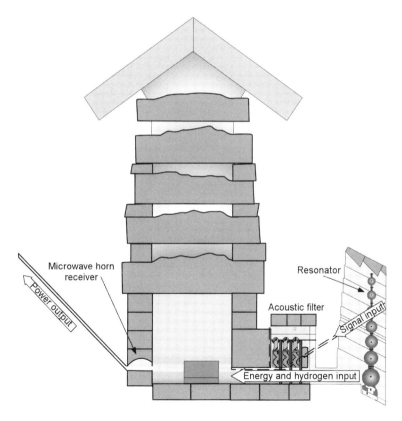

FIGURE 53. *The Power Center*

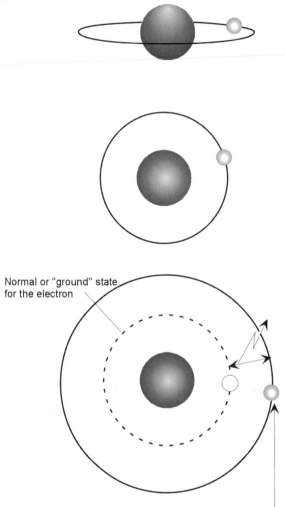

Normal or "ground" state
for the electron

The atom absorbs energy through resonance, and the electron is
"pumped" to a higher energy state. When the electron falls back to
ground state, it releases a packet of electromagnetic energy with a
frequency that is in the microwave region.

FIGURE 54. *Hydrogen*

With the granite beams vibrating at their resonant frequency, the sound
energy would be converted through the piezoelectric effect of the silicon-
quartz crystals embedded in the granite, creating high-frequency radio waves.
Ultrasonic radiation would also be generated by this assembly. The hydro-

gen generated in the Queen's Chamber, directly below the King's Chamber, would fill the upper chambers and then efficiently absorb this energy as each atom responded in resonance to its input.

Hydrogen, as I explained, is the simplest atom, having only one electron and one proton. The electron is "pumped" with energy to a higher energy level. In other words, the electron is induced to increase its distance from the proton. This is an unnatural state for the electron to be in, and in time it will fall back to its "ground state," releasing a packet of energy as it falls (see Figure 54). The electron can be stimulated to fall back to ground state through the action of an input signal—another packet of energy—that is of the same frequency. The result is that the input signal continues its path after stimulating emission from the hydrogen atom and carries the energy released by the electron with it.

In the Giza power plant, the Northern Shaft served as a waveguide through which the input microwave signal traveled. A typical waveguide is rectangular in shape, with its width being the wavelength of microwave energy and its height measuring approximately one-half its width. The Northern Shaft waveguide was constructed precisely to pass through the masonry from the north face of the pyramid and into the King's Chamber. That mi-

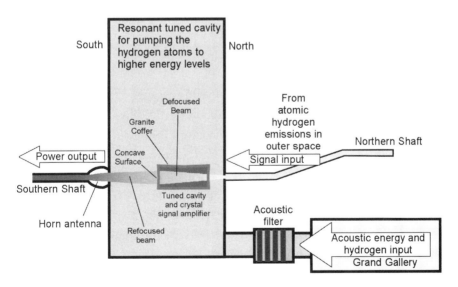

FIGURE 55. *The Pyramid MASER*

crowave signal could have been collected off the outer surface of the Great Pyramid and directed into the waveguide (see Figure 55).

The (originally smooth) surfaces on the outside of the Great Pyramid are "dish-shaped" and may have been treated to serve as a collector of radio waves in the microwave region that are constantly bombarding the Earth from the universe. Amazingly, this waveguide leading to the chamber has dimensions that closely approximate the wavelength of microwave energy—1,420,405,751.786 hertz (cycles per second). This is the frequency of energy emitted by atomic hydrogen in the universe. These features and facts are gathered in Table 3.

TABLE 3

DESCRIPTION	MEASURE	UNIT
Frequency of atomic hydrogen (hf)[3]	1,420,405,751.786	hertz
Speed of electromagnetic radiation (light) per second (c)	186,282 11,802,827,520	miles inches
Wavelength of hydrogen microwave energy (c ÷ h)	8.309	inches
Width of Northern Shaft leading to King's Chamber[4]	8.4	inches
Height of Northern Shaft leading to King's Chamber[5]	4.8	inches

These features inexorably move us to consider the purpose for the gold-plated iron that was discovered embedded in the limestone near the Southern Shaft. In order to have an efficient conduit for electromagnetic radiation, the entire lengths of the Northern and Southern Shafts would have to have been lined with this material, thereby making a very efficient conduit for the input signal and the power output (see Figure 56).

I have puzzled for a long time over the granite box inside the King's Chamber. This box, now located at the back of the chamber, is a critical component in this maser, and to imagine how it may have been used I am compelled to move it from its current position and place it between the waveguides in the north and south walls. There is evidence that leads me to

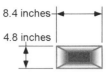

8.4 inches

4.8 inches

Opening in the
south wall

Opening in the
north wall

Microwave horn
antenna

Waveguide

The north wall in the King's Chamber

Evidence for the maser theory exists in the
form of a waveguide and a horn antenna.

FIGURE 56. *Microwave Horn Antenna and Waveguide*

suggest that it occupied this position and served to amplify the microwave signal that entered the resonant cavity. To understand how the granite box may have functioned, it would help to understand the basic principles of how optics function. Normally we associate optics with visible light. Most of us are familiar with telescopes, binoculars, and spectacles, and—with the exception of mirrors—naturally assume that we should be able to see through them. But that is not always the case. The material from which an optic is made depends on the wavelength of electromagnetic radiation that passes through it. As humans, we are equipped to see electromagnetic radiation (light) that lies in what we call the visible spectrum. There is light, however, below and above the visible spectrum that we are not physically equipped to see. If we were, we would be able to see through some of the opaque materials that allow these wavelengths to pass freely.

For example, the wavelength of a Nd.YAG (Neodymium.Yttrium-Arsenate-Garnet) laser is 1.06 microns. Optical components that pass light

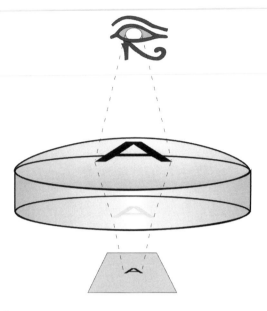

FIGURE 57. *Refraction*

of this frequency also pass visible light. A CO_2 laser, on the other hand, has a wavelength that is 10.6 microns—ten times greater than a Nd.YAG—and the most efficient and cheapest material for passing CO_2 laser light is gallium arsenide, which humans like you and I cannot see through. If we want an example of a material that is opaque and yet passes electromagnetic radiation, we have to look only as far as the kitchen cabinets for the containers that we place in the microwave oven. Microwaves pass through an opaque dish and heat the food it contains. The granite coffer, densely opaque to us, would allow electromagnetic radiation, invisible to us, to pass through.

Having a material present that allows microwave energy to pass through, we can apply the basic principles of optics that affect all wave phenomena, including electromagnetic radiation. These principles are reflection and refraction. Reflection, of course, is a look in the mirror. Refraction explains the focusing of a lens, such as a magnifying glass (see Figure 57). For instance, lenses are ground with a curvature that refracts light in such a way that it is either focused or diverged, depending on the application. Eyeglasses are ground so that they focus the light rays that pass through them, bringing objects into clearer view.

There is evidence to suggest that the granite box could refract electro-magnetic radiation as it passed through the box's north and south walls. Though fully accurate measurements for optical characteristics have not been made on these surfaces, Smyth's measurements show that the grinding on these surfaces produced a concave surface.[6] Manufactured in such a manner, the coffer—positioned in the path of the incoming signal through the North-ern Shaft and with oscillating crystals adding energy to the microwave beam—may have served to spread or diverge the signal inside the box as it passed through the first wall. Within the confines of the granite box, the spreading beam would then interact and stimulate the emission of energy from the energized, or "pumped," hydrogen atoms (see Figure 58).

If we follow a straight line across the King's Chamber from where the Northern Shaft enters, we find a feature cut into the granite wall that closely resembles a horn antenna, much like a microwave receiver. Passing through the opposite wall of the coffer, then, the radiation picked up more energy, was once more refracted, and then focused into this horn antenna. The mouth of this opening shows signs of being severely damaged. Because of the curved geometry of this opening, somebody in the distant past probably found it necessary to hack away some of the granite in order to retrieve the gold or gold-plated metal lining. Nevertheless, what is left unmistakably identifies this feature as the receiver of microwave energy that entered the chamber from the waveguide in the north wall.

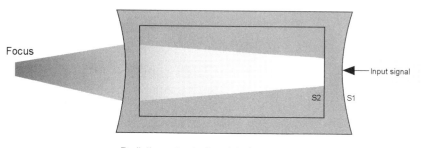

Radiation refracts (bends) when passing through one medium to another. Radiation is refracted when passing through both S1 and S2.

FIGURE 58. *Lens Effect of the Granite Box*

With the preceding data, we can explain many mysteries found in and around the King's Chamber. Compelling evidence clearly identifies the King's Chamber as the power center of the Giza power plant and strongly suggests that microwave energy was flowing through the southern "air shaft" and utilized on the outside. However, as I said earlier, any theory purporting to explain the purpose of the Great Pyramid should explain *all* noted phenomena. We still have not addressed the purpose for the Queen's Chamber, Horizontal Passage, Well Shaft, and Subterranean Pit. Neither have we studied how the hydrogen gas may have got inside the pyramid. As a matter of fact, we can answer all of these questions if we look more closely at the Queen's Chamber, where we will find the evidence that proves it was used to generate hydrogen, the fuel that ran the Giza power plant.

Chapter Eleven

A HYDROGEN GENERATOR

am indebted to Rudolph Gantenbrink for his exploration of the Southern Shaft in the Queen's Chamber. I can understand his frustration at not being allowed to explore further inside the Great Pyramid, but am delighted that his discovery at the end of the shaft provided conclusive evidence for what I am about to discuss in this chapter.

Without hydrogen this giant machine would not function. It was the medium by which the energy drawn through the Great Pyramid was converted and transmitted to the outside. The hydrogen for the operation of the maser was generated by a chemical reaction in the Queen's Chamber. The characteristics and discoveries in this chamber strongly suggest that two solutions—such as hydrated zinc chloride and a dilute solution of hydrochloric acid—may have been introduced to cause a chemical reaction that produced hydrogen. There are other ways to create hydrogen, such as electrolysis; however, I am going to discuss only one method and the evidence that exists to support it.

The Queen's Chamber is located in the center of the pyramid. There are two shafts leading to this chamber which were bored into the wall block and terminated five inches from the inside wall of the chamber, leaving what Smyth described as a "left" (see Figure 59). The discovery of these enigmatic shafts came in 1872 when Waynman Dixon was able to thrust a rod through a small crack in an otherwise perfectly fitted wall and then chiseled through the limestone. Dixon noted that the limestone "left" was particularly soft in that area. Emboldened by this important discovery he measured off the same distance on the other wall and discovered another shaft. The channels were quickly theorized to be air channels; and it was surmised that the builders were not required to ventilate this chamber at

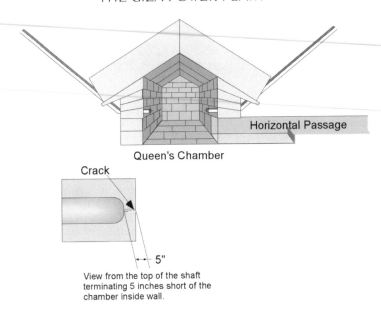

FIGURE 59. *Queen's Chamber Shafts*

the time of building, but were making preparations in case they might change their minds at a later date.

While some Egyptologists have proposed that the shafts leading to the Queen's Chamber were included in the design to provide air to that chamber, there is a simple and obvious fact that proves the theory to be wrong. A shaft cannot pass air from the outside to a chamber if the shaft is blocked at both ends—as the Queen's Chamber shafts are. Therefore it is amazing that such a speculation would be made. Even if these shafts were open at both ends, they do not have the most efficient design to be air shafts. Although it could be argued that the shafts were built on an incline to make the shortest route to the outside, a horizontal—albeit longer—shaft would be quicker and easier to build. For a horizontal shaft, the faces of the blocks and the channel that runs through each block, comprising the walls and ceiling of the shaft, would not have to be cut on an angle (see Figure 60). Neither would the surface that provided the floor for the shaft. The blocks could sit on a single course of masonry.

Smyth is credited for noting another anomaly in the Queen's Chamber—there were flakes of white mortar exuding from stone joints inside the shaft. Analysis of the mortar found it to be plaster of paris—gypsum (calcium sulfate). Smyth also described this chamber as having a foul odor, which caused early visitors to the chamber to beat a hasty retreat, and it was assumed that tourists were relieving themselves, though the way Smyth described this chamber, few people stayed long enough to do so. However, as I will make clear, this odor may not have been the result of unhygienic conditions but of the chemical process that once occurred in the Queen's Chamber.

One of the greatest mysteries of this chamber has been the salt encrustation on the walls. It was up to one-half-inch thick in places, and Petrie took it into account when he made measurements of the chamber. The salt also was found along the Horizontal Passage and in the lower portion of the Grand Gallery. How did salt come to build up on the walls?

Those who have seen some significance in the presence of the salt have speculated that it could have been deposited on the walls as the water of the biblical Great Flood receded. Others have speculated that the Great Pyramid and its neighbors were surrounded by water at one time. There has been no

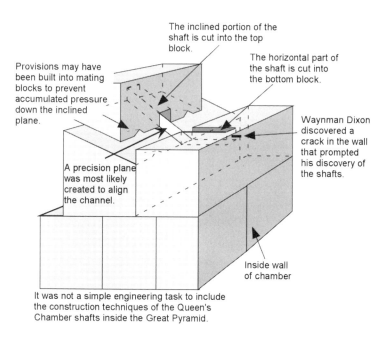

FIGURE 60. *Southern Shaft in the Queen's Chamber*

evidence, however, to support such speculations; in fact, one researcher, after clambering over the sides of the Great Pyramid looking for water marks, concluded that such speculations were groundless.

The flood and groundwater theories do not account for the salt found in the Great Pyramid, but proponents of these theories might be right about one thing—the existence of the salt itself suggests that the Queen's Chamber was designed to take in fluid rather than air. The power plant theory explains why fluid may have been introduced into the pyramid and how it produced the salt encrustations on the walls. Nearly all of the facts that have so mystified Egyptologists fit logically together if the Queen's Chamber was once used to create hydrogen to fuel this power plant.

Salt is a natural by-product of the reaction designed to produce hydrogen. It would form when the hot, hydrogen-bearing gas reacted with the calcium in the pyramid's limestone walls. If the half-inch of salt on the Queen's Chamber walls were the result of repeated inundations—not of water—of chemicals, those used to create the fuel that powered the plant, it is unlikely that any of this chemical fluid would find its way into the bottom part of the Grand Gallery. The orientation of the passage leading from the Queen's Chamber would take the chemicals down into the Well Shaft and into the Subterranean Pit. Because the greatest concentration of heat would be inside the Queen's Chamber, with a gradual cooling off as the gas made its way into the Grand Gallery, the buildup of salt would diminish along with the heat. This process would account for the presence of salt in the Queen's Chamber and other parts of the pyramid and also for its uneven distribution in these areas.

The kind of salt that is created in a chemical process depends on the chemicals involved in generating the gas. I am not going to argue for a particular combination of chemicals, and say that is the combination the ancient Egyptians used. But I will present one feasible combination—suggested to me by a chemical engineer—with the understanding that the pyramid builders may have used a different one. Before I proceed, however, we must consider the material the pyramid builders used to bind those limestone blocks adjacent to the shafts in the Queen's Chamber. We remember that Smyth noted a white, flaky material exuding from the joints of these blocks. He was sufficiently interested in this material to have it chemically analyzed, and it was found to be gypsum, or plaster of paris. However, it is possible that this material was not used by the builders in this area at all, and if not, then we are provided with another clue to the real purpose of the Queen's Chamber. Let us briefly detour in order to find out a little more about exactly what plaster of paris is and how it could have come to be in the shafts of the Queen's Chamber.

Plaster of paris is dehydrated gypsum, which, when moistened with water, sets in a relatively short time. It got its name because gypsum was widely used near Paris to make plaster and cement. In nature, gypsum is a form of sedimentary rock, formed by the precipitation of calcium sulfate from seawater. It also is found in various other saline deposits.

It is possible that the ancient Egyptian builders used plaster of paris in the joints in the shafts to retard the absorption of fluid into the surrounding masonry. However we should not exclude other possibilities—one of which is another method of creating gypsum that could lead to some extremely interesting conclusions. Gypsum also is produced through the action of sulfuric acid on limestone, and, although this in itself does not prove that sulfur and/or its by-products were used in the chemical process in this chamber, it does promote the consideration of other data in a new light. Because the pyramid's building materials contain one of the elements needed to produce gypsum (the limestone masonry), it follows that the introduction—either accidental or purposeful—of another necessary element (sulfuric acid, for instance) would produce gypsum.

Several questions come to mind in light of the preceding speculations. They may be totally unrelated, but it would be interesting to know the answers to the following:

- Was the disgusting smell that caused early explorers to beat a hasty retreat from the chamber connected to a chemical process that used sulfur? Hydrogen sulfide is particularly odorous, exuding a smell similar to rotten eggs. This gas is formed by the combination of sulfur with hydrogen. While early explorers expected to be confronted with a certain amount of bat dung inside the pyramid, it seemed as though the smell in this particular chamber was more pronounced than in the rest of the pyramid. Again, the composition of the salts on the chamber's walls may help clarify this investigation, as sulfur-bearing compounds may have formed these salts.

- Where did Caviglia get the chunks of sulfur that he burned in the Well Shaft? While it was a practice of early explorers to burn sulfur to purify unhealthy air, it would be most helpful to know whether or not he had the sulfur with him or if it was already there.
- If a chemical exchange process was separating hydrogen and if a catalyst was being used in the Queen's Chamber, would sulfur play a part in the operation or perhaps regeneration of the catalyst?[1]

The shafts leading to the Queen's Chamber revealed other oddities that may help our investigation into its true function. In these channels, explorers found a small bronze grapnel hook, a piece of wood, and a stone ball. Though their discovery was much publicized at the time, Egyptologists say little about these artifacts today; and if Graham Hancock and Robert Bauval had not showed an interest, the experts would probably have preferred to leave it at that. Hancock and Bauval took it upon themselves to search for these relics. They contacted Egyptologist I.E.S. Edwards and the British Museum in 1993, but were told that they had no knowledge of them. The *Independent*, a British national newspaper, published an interview with Edwards in which he categorically denied any knowledge of these artifacts. Surprisingly, the following week, Dr. Vivian Davies, the Keeper of Egyptian Antiquities at the British Museum, stated in a letter to the *Independent* that the relics were in a cigar box in his department's keep. Subsequent investigation by Hancock and Bauval revealed that in 1972, Edwards had received the relics and recorded them in the museum's log.[2]

Was the fiasco surrounding the search for these relics a deliberate deception by Egyptologists? I seriously doubt it. These relics are pieces of evidence that do not fit into the orthodox tomb theory, so they probably had not been given any thought since the day they were delivered to the museum. Even if it was deceit—and the museum had intentionally kept them hidden—having the relics now on display does not answer the question of why they were in the pyramid in the first place. Several questions remain unanswered regarding the discovery of these items, and the items themselves, that prevent any assertion as to their intended purpose:

- Was there just one of each item?
- Were they found together in the same shaft?
- Were they in any way connected?
- Was their removal from the shaft difficult or easy?

If we work on the premise that the items were discovered in the same shaft and were connected in some way, or were able to be joined together, we could theorize that they might have been part of the switching mechanism that signaled the need for more chemicals. This hypothesis makes sense when viewed in context with other evidence from the Queen's Chamber. When we review the details researchers have noted inside the Queen's Chamber, all these seemingly unexplained facts fall into place, if our explanation is founded on the basic premise that some kind of chemical reaction was taking place there.

There is a corbelled niche with a small tunnel cut to a depth of thirty-eight feet, ending in a bulb-shaped cavern. Hydraulics engineer Edward Kunkle (of Ohio and now deceased) questioned the official explanation that this tunnel was cut by treasure seekers and claimed instead that with its flat, level floor and its almost perfect right-angled left side, it must have been part of the original construction. What is more, in Kunkle's view its features show it could have served a mechanical purpose. Kunkle proposed that it was part of a large ram pump, which also involved other features located inside the Great Pyramid.

In his book *5/5/2000 Ice: The Ultimate Disaster,* Richard Noone focused upon the mysteries of this chamber and asked a tantalizing question. Noone went further in his search for reasons for the presence of salt in this chamber. He interviewed Dr. William Tiller of Stanford University, whose research deals with the science of crystallization, surfaces, and interphases between two media, a science concerning biomaterials and psychoenergetics. Other than an interesting comment on "energy resonances" that he says would be particularly good to excite things in a pyramid, Tiller could not shed any light on the cause of the mysterious salt encrustations. Noone, however, had an enlightening dialogue with him:

Noone: Dr. Tiller, why would the Middle Chamber of the Great Pyramid be such a magnet, or repository for these crystals?

Tiller: That's really tough to imagine, because you basically have got
to have (a) either some salt water which was in there and evaporated and then deposited this because it got very hot, or (b) it
was at some time under the ocean and the water seeped in, or
(c) somehow the "energies" converted what was in the walls to
sodium chloride. None of these, however, makes any sense at all
to me. I regret I don't have, at this point in any event, a really
good answer for you.[3]

I was hoping to be able to get into the Queen's Chamber while I was in
Egypt in 1986 to get a sample of the salt for analysis. I had speculated that
the salt on the walls of the chamber was an unwanted, though significant,
residual substance caused by a chemical reaction where hot hydrogen reacted with the limestone. Unfortunately, I was unable to get into the chamber because a French team was already inside the Horizontal Passage, boring
holes into what they hoped were additional chambers. (It was discovered,
after I left Egypt, that the spaces contained only sand.)

As it turned out, my research would have been redundant. Noone reported in his book that another individual had already had the same idea and
done the work. In 1978, Dr. Patrick Flanagan asked the Arizona Bureau of
Geology and Mineral Technology to analyze a sample of this salt. They found
it to be a mixture of calcium carbonate (limestone), sodium chloride (halite or
salt), and calcium sulfate (gypsum, also known as plaster of paris). These are
precisely the minerals that would be produced by the reaction of hot, hydrogen-bearing gas with the limestone walls and ceiling of the Queen's Chamber.

Armed with this information, I sought out a chemical engineer, Joseph
Drejewski, to see if my perception regarding the Queen's Chamber was plausible. He was skeptical about my entire premise but agreed to look at the
data and form an evaluation.

I had speculated that the five-inch "left"—which contained a small
hole through to the channel—that prevented each shaft from joining with
the Queen's Chamber was intentionally designed to meter a specific amount
of fluid into the chamber over a period of time. If we knew the head pressure of the fluid, we could accurately calculate the amount of fluid that
flowed through this "left". One of these two shafts has a different discolora-

tion, or staining, and I speculated that this was the result of the ancient Egyptians introducing two different chemicals into the chamber, which, when combined, would produce a reaction. Drejewski agreed that two chemical solutions could be introduced into this chamber to create hydrogen or ammonia under ambient conditions of 80° Fahrenheit, ± 20°. He agreed that the niche in the wall of the chamber could have been used to house a cooling or evaporation tower. The corbelled niche inside the reaction chamber would have provided an anchor for this tower, which also may have contained a catalyst (see Figure 61). One scenario could be that the chemicals pooled on the floor of the chamber and wicked through the catalyst material. The offset of the niche may indicate the proportion of each chemical introduced into the chamber. Drejewski, therefore, agreed that my theory was plausible.

To evaluate my theory further, we must now move from considering the technology of this machine to the fuel that ran it. Let us consider how hydrogen is made and used to produce energy (see Figure 62). Drejewski prepared a report informing me of the following:

Hydrogen is most easily obtained by displacing it from acids by con-

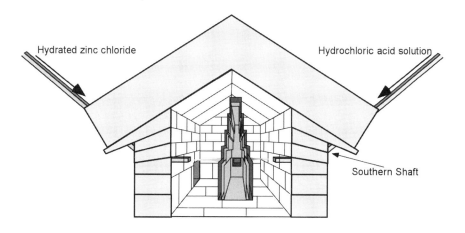

The corbelled niche inside the Queen's Chamber
equipped with cooling/evaporator tower.

FIGURE 61. *Cooling/Evaporator Tower*

tact with certain metals that are more active than hydrogen and there-fore will combine more readily with the other constituents of the acid. Zinc (Zn) is the most commonly chosen metal and when treated with dilute hydrochloric acid (HCl), it will produce a reasonably pure hy-drogen gas which evolves at a relatively fast rate. The hydrogen gas produced by this reaction of zinc with hydrochloric acid may contain water vapor carried along by the gas as it bubbles through the water solution. If impurities are present, it is possible to remove the water vapor (with the impurity) by passing the generated gas over or through a drying agent such as calcium chloride (Ca Cl$_2$), which retains the water vapor, but does not react with the hydrogen gas. Other metals which can be used [as a drying agent] are magnesium and finely divided iron (powder).[4]

Drejewski ended his evaluation by cautiously stating, "It is highly prob-able that [through this reaction] impurities such as calcium sulfate (gyp-sum) and sodium chloride (halite) can be leached through limestone [cal-cium carbonate] (Ca CO$_3$)."

I have been asked by other researchers who have reviewed a synopsis of my theory whether electrolysis could have played a part in the generation of hydrogen. I am not going to rule that out completely, but electrolysis would require only one shaft leading to the chamber, as it is a process using only water and electrical power. We have to explain the reasons for two similar shafts, and the dark staining inside the Northern Shaft. This staining clearly indicates the use of two different chemicals.

Additional evidence to support the theory that chemicals were flowing

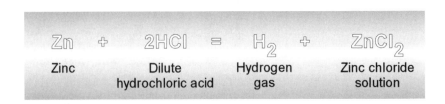

$$Zn \ + \ 2HCl \ = \ H_2 \ + \ ZnCl_2$$

| Zinc | Dilute hydrochloric acid | Hydrogen gas | Zinc chloride solution |

FIGURE 62. *The Chemical Process to Produce Hydrogen*

down these shafts came in 1993, when Rudolph Gantenbrink guided his robot Upuaut II up the Southern Shaft and discovered the so-called "door" with its copper fittings. We will remember that Smyth noted gypsum exuding from the joints of the Southern Shaft leading to the Queen's Chamber. The filming of this channel by Gantenbrink's robot revealed signs of erosion in the lower portion of the shaft. The walls and floor of the channel were extremely rough, and the erosion of the walls appeared to have horizontal striations. There also were signs of what appeared to be gypsum leaching from the limestone walls. Having reached an opinion regarding the function of the Queen's Chamber in the Great Pyramid, I was quite intrigued when the discoveries in this channel were publicized. I did not know if my speculations would be bolstered by what was found there or if I would have to throw them out. As it happened, my theorized function of the Queen's Chamber within the Giza power plant was strengthened.

I have proposed that the Queen's Chamber was designed to allow the necessary chemical elements into the chamber at a metered rate. However, considering that the limestone masonry blocks inside the Queen's Chamber had perfectly fitted joints, we could reasonably ask if the crack in the wall was really an anomaly or if, instead, it was part of the original design. In the context of my theory, and if so designed, this "left" with the crack in it could have served to meter the chemicals into the chamber.

Gantenbrink's robot came to a dead end at the upper part of the Southern Shaft. It encountered a block of limestone with two mysterious copper fittings protruding through it. It was widely publicized that a hidden door had been found inside the Great Pyramid (see Figure 63). What was not publicized, however, is that the shaft itself is only about nine inches square. The so-called "door," I believe, is a misnomer. As for the copper fittings—in the documentary they are presented as being stops to prevent the limestone block in which they are located from being raised. But this explanation does not fit. Why would the pyramid builders want to include a sliding block in an inaccessible area? And even if they did, how was it activated?

While I was watching the video of the exploration with my friend, Jeff Summers, he off-handedly remarked that the fittings looked like electrodes. His observation made sense to me. To deliver an accurate measure of hydrochloric acid solution to the reaction chamber, a certain head pressure would

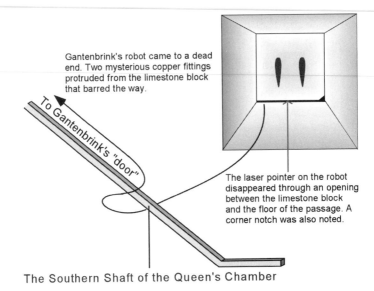

Gantenbrink's robot came to a dead end. Two mysterious copper fittings protruded from the limestone block that barred the way.

To Gantenbrink's "door"

The laser pointer on the robot disappeared through an opening between the limestone block and the floor of the passage. A corner notch was also noted.

The Southern Shaft of the Queen's Chamber

FIGURE 63. *Gantenbrink's "Door"*

need to be maintained. The head pressure is determined by the volume of fluid in the channel, that is, the weight of the column of chemical. The copper fittings would have served as a switch to signal the need for more chemicals. Floating on the surface of the fluid would have been another part of this switch—the cedarlike wood joined together with the bronze grapnel hook. This assembly would rise and fall with the fluid in the channel. With the channel full, the bronze prongs would have made contact with the electrodes, creating a circuit, and as the fluid in the channel dropped, the prongs would move away from the electrodes, thereby breaking the contact and acting as a switch to signal the pumping of more chemical solution into the channel until the bronze hook again made contact with the electrodes (see Figure 64). As the rate of supply into the reaction chamber was slight, a small opening was all that was needed to maintain the supply of chemicals. The limestone door with its copper fittings, brought into the view of the camera when Upuaut II had gone as far as it could, has a slight gap at the bottom, under which the robot's laser light disappeared. There also is a notch in the bottom right-hand corner. All these features serve to support the speculation that the Egyptians were supplying a fluid to the Queen's Chamber

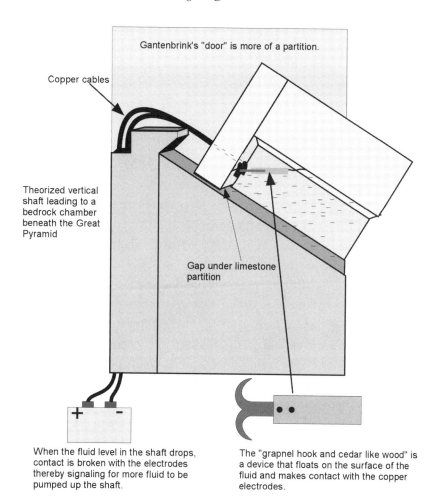

Gantenbrink's "door" is more of a partition.

Copper cables

Theorized vertical shaft leading to a bedrock chamber beneath the Great Pyramid

Gap under limestone partition

When the fluid level in the shaft drops, contact is broken with the electrodes thereby signaling for more fluid to be pumped up the shaft.

The "grapnel hook and cedar like wood" is a device that floats on the surface of the fluid and makes contact with the copper electrodes.

FIGURE 64. *Fluid Switch*

shafts, and it was necessary to maintain the fluid level in the shafts so that the weight of the fluid assured a constant and precise flow through the "left" in the chamber wall.

Where the robot ended its trek up the Southern Shaft, the walls, ceiling, and floor of the shaft were smooth, as they would have been when first fitted together. On the other hand, the texture of the walls and floor of the shaft at the lower level—as photographed by Gantenbrink—was deeply eroded, with horizontal striations, and there also appeared to be leaching of salts on the surface of the ceiling and the walls. Both of these conditions could have been caused by a chemical fluid.

The recent investigation by Tom Danley in the Great Pyramid took him into the Queen's Chamber, where he tested the Southern Shaft using an acoustic device. By calculating the time it took for the sound wave to travel the distance of the shaft, bounce off the "door," and return to the sensing device, he was able to determine the length of the shaft. But Danley detected something else. The device heard a secondary echo. This echo was produced by the sound squeezing through the small gap at the bottom of the door and traveling through into the space beyond, the space where Gantenbrink had offered to go with another robot but was denied the opportunity. Danley's instrument told him that the sound wave traveled another thirty feet before bouncing back to the source.

Notwithstanding the fact that there is no tangible evidence of what lies behind Gantenbrink's "door," what has been discovered fits extremely well into the power plant theory. It also allows us to extrapolate, for the present, what lies hidden from view and predict what further explorations may uncover when the block Gantenbrink discovered is finally penetrated. A hint of what might be there came in 1992 when French engineer and professor Jean Leherou Kerisel, and his team, conducted ground-penetrating radar and microgravimetry tests in the short horizontal passage that leads from the Descending Passage to the Subterranean Pit. He detected a structure under the floor of the passageway which he analyzed as possibly being a corridor oriented SSE–NNW and with a ceiling at the same depth that the Descending Passage would have reached had it been continued.

Hancock and Bauval reported on Kerisel's findings:

Nor was this all. A second very clear anomaly, a "mass defect" as Kerisel calls it, "was detected on the western side of the passageway six metres before the chamber entrance. According to our calculations, this anomaly corresponds to a vertical shaft at least five metres deep with a section of about 1.40 x 1.40 metres very close to the western wall of the passageway."

In short, what Kerisel believes he has identified off the Subterranean Chamber's entrance corridor is something that looks very much

like a completely separate passageway system, terminating in a verti-
cal shaft. His instruments may have misled him, or, as he himself
admits, he may merely have picked up the traces of "a large volume of
limestone dissolved by the action of underground water—in other
words a deep cave". Alternatively, however, if the "mass defect" turns
out to be a man-made feature, as he strongly suspects, then "it may
lead to very interesting developments."[5]

Kerisel's findings indicate that the supply shafts leading to the Queen's Chamber may have been supplied with chemicals by means of a vertical shaft that connected to an underground chamber. It should be noted that Kerisel detected the vertical anomaly on the west side of the passageway. The shafts leading to the Queen's Chamber are oriented to the west of the passageway. In light of my proposed use for these shafts, and of Kerisel's discovery, it would not be out of order for us to postulate that when Gantenbrink's "door" is penetrated, or when the clandestine diggers above the King's Chamber reach their destination, a vertical shaft leading to a bedrock chamber will be found. I also would not be surprised if more copper, in the form of cables or wires that had been attached to the "copper fittings," are found beyond Gantenbrink's "door" (refer to Figure 64).

We can now understand how chemicals were introduced to the Queen's Chamber and caused a reaction that filled all the cavities within the Great Pyramid with hydrogen. But during the fueling process, moisture and impurities would had to have been removed from the gas. How was this done? The means existed to remove water vapor and impurities from hydrogen gas. The gas had already given up some moisture and impurities in the Queen's Chamber, leaving salt encrustation on the walls and ceiling. The long Horizontal Passage that connects the Queen's Chamber with the Grand Gallery, being constructed out of limestone—the same material as the Queen's Chamber—would work in the same manner and remove residual moisture and impurities from the gas as it filled that passage and flowed toward the Grand Gallery.

At the juncture where the Horizontal Passage meets the Ascending Passage is a five-inch lip. There may have been a slab resting against this lip and bridged between the Ascending Passage and the floor of the Grand Gallery, where a similar lip is found. Slots in the sidewall indicate that there may have been supporting members for this slab, which would have had holes drilled into it to allow the gas to rise into the Grand Gallery. At this juncture, and to the west, a hole in the wall leads to what is known as the Well Shaft. Perhaps the Well Shaft is the only feature of the Great Pyramid that has been accurately named. Spent chemical solution from the Queen's Chamber would have flowed along the Horizontal Passage and down the Well Shaft into either the Grotto or, if the shaft at that time connected to the lower Descending Passage, the Subterranean Pit below. The lip and a bridging slab would have prevented the fluid from flowing down the Ascending Passage (see Figure 65).

One of the questions that always arises regarding my theory about the Great Pyramid, and one that should be on our minds at present, is what happened to the original equipment that sustained the operation of the power plant? While it is a generally accepted belief that some objects have been removed from the Great Pyramid, it is impossible to define exactly what was removed, when it was removed, and who removed it. Tradition teaches us that it was a king's mummy and vast treasures that made grave robbing such a lucrative business in Egypt. However, considering the proposition in these pages, other treasures removed from the Great Pyramid were far more valuable than the funerary trappings that accompanied an Egyptian king. Why these objects were removed and by whom I cannot say, but the features we examined from the King's and Queen's chambers, and from the Grand Gallery and its juncture with the Horizontal Passage, lend themselves to speculation that objects were there and provide us a clue to what these objects

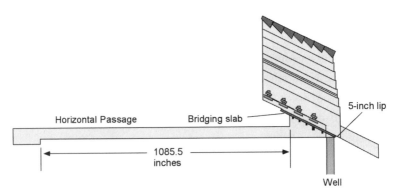

FIGURE 65. *Bridging Slab*

were. As I have just argued, I believe these objects were parts of a great machine, the Giza power plant. The evidence in the Queen's Chamber (see Figure 66) alone points to a finely tuned operation that used a chemical reaction to create hydrogen. Let us review this evidence.

- *five-inch lip at juncture of Ascending Passage and Horizontal Passage.* This could serve to prevent waste from flowing down the Ascending Passage.
- *drop in Horizontal Passage floor level before it goes into the Queen's Chamber.* This would allow chemicals to pool inside the chamber and "wick" up the evaporation tower (refer to Figure 62).
- *corbeled niche in wall of the Queen's Chamber.* This may have been a means to "key" the evaporation tower into the structure.
- *shafts leading to the Queen's Chamber but not quite connected to it.* These could have been supply shafts for chemicals needed in the reaction. The shafts would allow chemicals to enter the chamber and prevent evolving gases from escaping.
- *stone ball, grapnel hook, and cedarlike wood.* The wood and hook assembly could have served as a floating contact to signal the need for more chemical. The stone ball may have been used to prevent erosion of the "left" as the channel filled with fluid.
- *flakes of gypsum exuding from joints in shafts.* This substance probably resulted from the chemical reacting with limestone (suggesting the use of sulfuric acid).
- *buildup of salt crystals on the walls and ceiling of the Queen's Chamber, Horizontal Passage, and lower level of Grand Gallery.* This buildup was likely the result of gaseous vapor passing over the limestone, reacting with the calcium in the limestone, and giving up water and impurities. This was a by-product from the drying of the gas.
- *Well Shaft bored from the juncture of the Grand Gallery and the Horizontal Passage down to the Grotto.* This was probably either a waste removal shaft or an overflow shaft.
- *large granite block at the bottom of the Well Shaft at the level of the Grotto.* Most likely this was put into place to catch the chemical overflow, thereby preventing erosion of the limestone.

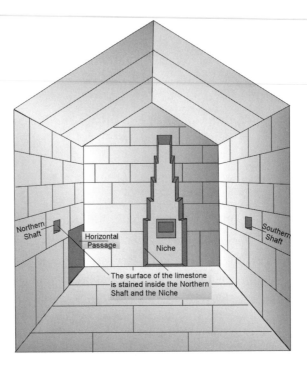

FIGURE 66. *Niche Inside the Queen's Chamber*

It is unfortunate that such havoc was wreaked inside the Great Pyramid as early explorers bored and tore away barriers. The Descending Passage and the Well Shaft were recipients for much of the refuse from those expeditions. More recently another area of the pyramid has allegedly been used as a waste disposal site for limestone residue from tunneling. There are speculations that there is a clandestine effort to reach Gantenbrink's "door" by tunneling from one of the chambers above the King's Chamber. Tom Danley reported that the limestone chippings from this effort were being placed in burlap bags and taken to another chamber above in an effort to keep the project a secret. If this is true, we have reason to worry, for if the tunnelers ultimately reach their destination, who knows what damage they will inflict in this area? In a manner similar to explorations of earlier times, valuable features may be destroyed—by the pickaxes and chisels of these modern treasure seekers—because they are unrecognizable within the context of a tomb.

Chapter Twelve

MELTDOWN

 *uch of the evidence that supports the theory that the Great Pyra-mid* was a power plant is the result of a malfunction of the gen-eration process. The hydrogen in the power center (King's Cham-ber) for some inexplicable reason exploded in an awesome ball of fire, and the power plant suffered a "meltdown." The King's Chamber was affected in a disastrous way. Its walls were pushed out nearly an inch and the ceiling beams cracked.

The operators of the power plant, noting that there was an interrup-tion of the energy coming from the pyramid, had to enter the pyramid to make repairs. They spread plaster over cracks in the ceiling beams, appar-ently making no pretense to neatness, for the plaster was daubed on freely, almost as though they had used their fingers. The question we must ask is, Would the spreading of plaster on these giant monoliths improve the struc-tural stability of the ceiling? Or was it for another reason that the Egyptians took the trouble to seal the cracks? It would seem that if the granite beams in the ceiling of the King's Chamber were suddenly to give way, a smear of plaster would do little to prevent them from smashing into the chamber. But if the Great Pyramid were a machine—a power plant—then the sealing of these cracks may make sense.

If the fuel that fed the operation of the power plant was hydrogen gas, it is conceivable that it was necessary for the operators to contain the gas as completely as possible—for an excessive leak might have decreased the effi-ciency of the power plant. Therefore, when the guardians entered the King's Chamber and found severe cracks in the ceiling, it may have made sense to them to seal the cracks while they were checking out the rest of the damage and making whatever repairs they thought necessary. Whether the sealing of the cracks was essential was probably a chance the guardians could not

THE GIZA POWER PLANT

afford to take; after leaving the pyramid and resuming the operating cycle, they would not have wanted to reenter the pyramid in order to repair a simple leak.

The structural displacement of the King's Chamber is not the only clue that the heart of the Great Pyramid experienced a powerful release of energy. Another supporting clue which may have been the result of the same events that caused the disturbance in the King's Chamber, was found in one of the so-called "relieving" chambers above. As we recall, when Howard-Vyse's men blasted through tons of limestone and granite and discovered the four chambers above Davison's Chamber, the first part they went into had a strange effect on them. They crawled out of the air space covered from head to toe with a fine, thin black powder. The floor of the chamber was covered with it. Analysis of the powder showed it to be exuviae, the cast-off shells and skins of insects.

This discovery has remained a mystery. Where did the exuviae come from? None of the other chambers contained it. The space directly above the King's Chamber contained nothing but bat dung. There were no living insects found in the Great Pyramid, and it is doubtful that a group of insects would single out this one chamber and collectively, or repeatedly over a period of time, shed their skins.

My theory may account for this black powder. The anomalous creation of energy within the King's Chamber, which forced the granite walls away from their original position and cracked the granite beams above, also may have been responsible for the exuviae in the chamber above Davison's Chamber. Insect shells are comprised mainly of calcium carbonate, and if we look for a source for calcium carbonate in the area, we find it in the core limestone masonry itself. The core blocks of the pyramid are comprised mainly of nummulitic limestone made up of fossilized seashells and foraminifers. If there was an explosion inside the King's Chamber of sufficient magnitude to push aside hundreds of tons of granite, it is possible that with that explosion, and in the presence of elevated temperatures, the surface layers of limestone in close proximity would be affected too. The scenario may have gone like this: The initial explosion jolted the entire granite complex, pushing the walls out and lifting the ceiling beams up off their support blocks. As the ceiling beams came crashing back down, they cracked along the south end and, at the same time, some of the limestone core masonry in the spaces above the King's Chamber may have been impacted and crushed by their fall, causing limestone dust to fill the air. As the crushed limestone hung in the air it quite literally could have cooked in the elevated temperatures of the hydrogen explosion and the fire that followed. The black calcium carbonate dust would have settled finally onto the tops of the granite beams.

The guardians, alarmed at the sudden malfunction within the power plant, then gained access to the inner chambers of the pyramid by climbing down the Descending Passage and up the Well Shaft to the level of the Grand Gallery. They cut through to what is now known as Davison's Chamber, where they inspected the next layer of granite. While in this chamber they could have cleaned away the limestone dust (exuviae) from the top of the beams, which is why the exuviae was not discovered until an opening was made by Howard-Vyse into the chamber above.

Another feature inside the granite complex known as the King's Chamber that is left unexplained by orthodox theories is the so-called sarcophagus. We have already discussed a purpose for this box, but we really have not addressed why the pyramid builders selected a type of granite for the box that was a different color than the granite with which they constructed the chamber. The box is chocolate colored, and there is no granite like that to be found in Egypt! It has been speculated that it came from the Americas or the mythical Atlantis. If this is true, why would the pyramid builders find it necessary to import a single, large block of chocolate-colored granite from across the world to construct a sarcophagus when they could have used red granite, of which there was plenty available in their own country?

Well, perhaps they did not.

Again my theory of the Giza power plant provides a reasonable answer. Perhaps the coffer was originally red, quarried at the same time, in the same place, as the rest of the granite used to construct the King's Chamber. If an object like this box was subjected to excessive energy levels, what would be the effects? Depending on other elements that were present at the time of the malfunction of the power plant, it is conceivable that certain changes would be recorded in any object fortunate enough to survive the accident. The comparatively thin sides and base of the coffer would naturally be more susceptible to excessive energy levels than would be the huge granite blocks comprising the walls and ceiling of the King's Chamber. It is possible, therefore, that the granite box, because of its thinner construction, did not have the ability to conduct the heat to which it was subjected and so it simply overcooked, causing the color change. Architect Jim Hagan, who is an expert in the application of stonework in construction, explained to me that the interior chambers of the Great Pyramid have the appearance of being subjected to extreme temperatures; and he claimed that the broken corner on the granite box shows signs of being melted, rather than simply being chipped away.

The awesome force unleashed inside the King's Chamber—of such magnitude that it melted granite—also would have consumed other susceptible materials. If the resonators in the Grand Gallery were made of combustible material, such as wood, they most likely would have been destroyed at the same time. Evidence to support this speculation comes from reports that the limestone walls in the Grand Gallery were subjected to heat and, as a result, the limestone blocks calcinated or burned. The disaster that struck the King's Chamber, therefore, may have been responsible for destroying the resonators.

After shuddering to a stillness that they had not experienced for years, or even decades, the inner chambers of the Giza power plant lay in smoking ruins. Not knowing to what extent their machine was damaged, the operators would choose a route that would leave the interior chambers intact, in case they were able to make repairs and put the power plant back in service. The most obvious route would have been the Well Shaft. As we now turn our attention to the Well Shaft, a feature around which much debate has swirled, its existence begins to make sense in the context of my theory. Many Egyptologists credit the guardians of the Great Pyramid with carving an access tunnel—the Well Shaft—to the inside of the Great Pyramid to inspect damage from an "earthquake." This is a reasonable assumption because only those with knowledge of the internal passageways of the Great Pyramid would know where to dig their tunnel. However, considering the meandering and tortuous path of the Well Shaft, it was either a remarkable stroke of luck that their bore came out where it did, or they were in possession of some fairly advanced instruments for detecting the location of the

Grand Gallery, or a connection between the Well Shaft and Grand Gallery already existed.

Petrie presented evidence that showed the Well Shaft, from the Grand Gallery to the level of the Grotto, to be part of the original structure of the Great Pyramid. Therefore, instead of having to carve out the entire length, the guardians only had to carve their hole to the level of the Grotto. When the pyramid was being built, the portion of the well that was bored through the bedrock down to the Grotto would have been accessible to the workers. Once the work was accomplished, however, the constructed portion from the bedrock up to the level of the Grand Gallery would only need to be large enough to allow passage of the chemical overflow from the Queen's Chamber. And this is exactly what we find when we examine the Well Shaft dimensions. Petrie described the constructed portion of the Well Shaft as being rather shoddy work. Furthermore, he could not understand why a block that was used in its construction was positioned a mere 5.3 inches from the Grand Gallery wall. He wrote:

> On examining the shaft, it is found to be irregularly tortuous through the masonry, and without any arrangement of the blocks to suit it; while in more than one place a corner of a block may be seen left in the irregular curved side of the shaft, all the rest of the block having disappeared in cutting the shaft. This is a conclusive point, since it would never have been so built at first. A similar feature is at the mouth of the passage, in the gallery. Here the sides of the mouth are very well cut, quite as good work as the dressing of the gallery walls; but on the S. side there is a vertical joint in the gallery side, only 5.3 inches from the mouth. Now, great care is always taken in the Pyramid to put large stones at a corner, and it is quite inconceivable that a Pyramid builder would put a mere slip 5.3 thick beside the opening to a passage. It evidently shows that the passage mouth was cut out after the building was finished in that part. It is clear, then, that the whole of this shaft is an additional feature to the first plan.[1]

Based on all the evidence, the only explanation for the constructed portion of the Well Shaft near the Grand Gallery is that it was enlarged to

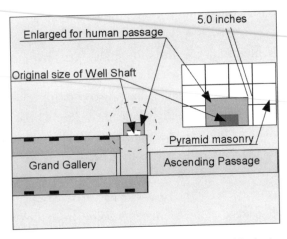

Petrie puzzled over the 5-inch thick limestone block at the mouth of the well, as it was not typical of the robust construction found throughout the rest of the pyramid.

Figure 67. *Well Mouth*

allow the guardians access to the Grand Gallery after the explosion in the King's Chamber, and this excavation resulted in the thin block of limestone on the south side (see Figure 67).

Evidence that the Well Shaft was part of the original design also was proposed by Celeste Maragioglio and Vito Rinaldi, who noted that the the walls upward from the Grotto to the Grand Gallery were lined with regular blocks of limestone. They argued that because a part of this lining was through the bedrock, it must have been a part of the original design and construction.[2] The existence of a large block of granite wedged in the mouth of the Grotto provides other evidence that the Well Shaft was part of the original design. The granite, being more impervious to erosion than the limestone, may have served to catch the chemical flow from the Queen's Chamber and direct it into the deep hole to the side of the Well Shaft. If the Well Shaft did not exist until the guardians made their damage assessment inspection, then they must have taken the block of granite from somewhere inside the Great Pyramid's main passages or chambers and dropped it down the Well Shaft. It makes more sense that the granite was a part of the original design, already in place, and that the guardians, or any other theorized interlopers, only had to push it aside when they reached the level of the Grotto (see Figure 68).

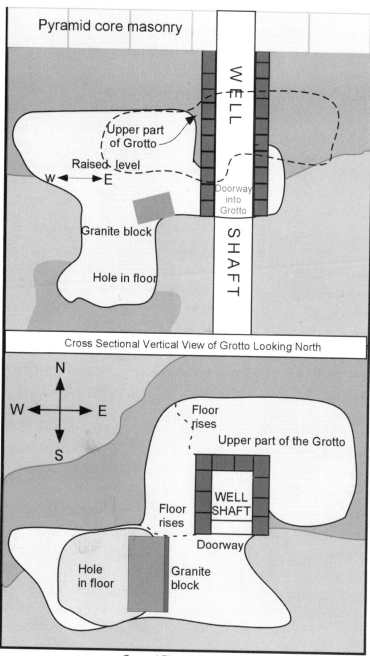

Pyramid core masonry

W E L L

Upper part
of Grotto

Raised level
W ← → E

Doorway
into
Grotto

Granite block

Hole in floor

S H A F T

Cross Sectional Vertical View of Grotto Looking North

N

W ← → E

S

Floor
rises

Upper part of the Grotto

WELL
SHAFT

Floor
rises

Doorway

Hole
in floor

Granite
block

Ground Plan of the Grotto

FIGURE 68. *Grotto in the Well*

The technology utilized in the Giza power plant was unique, and its design features find no parallel in any other structure anywhere in the world. Nevertheless, new technology does not just spring into existence. It is logical to assume that, before being encased in a mountain of stone, critical technological "devices," such as the King's Chamber granite complex, Grand Gallery resonators, and Antechamber acoustic filter, were all fully developed and successfully tested to the point where building the Great Pyramid became feasible. If we were to look for evidence that the ancient Egyptians undertook such development and testing, we need look no further than one-hundred yards to the east of the Great Pyramid, where the Trial Passages are located (refer to Figure 4).

These passages, discovered by Petrie and discussed earlier in the book, include features found inside the Great Pyramid suggesting that they were planned before the pyramid's construction. However, the ancient Egyptians did not excavate the Trial Passages out of solid bedrock just to demonstrate that they knew the Great Pyramid's interior design or for "practice," as many Egyptologists propose. They had a more practical purpose for their hard work. This purpose—indeed the very existence of the Trial Passages—becomes perfectly logical when considered within the context of the power plant theory. The Trial Passages on the Giza Plateau were most likely dug to accommodate the lower parts of the equipment being developed. As with many industrial and scientific research facilities around the world, the Trial Passages were dug to economize on the superstructure, which, in all probability would have been a research and development laboratory. Similarly today, industries installing large pieces of equipment save money by digging pits and lining them with concrete instead of raising the roof. Normally, the shape of these concrete pits is similar to the shape of the equipment they will accommodate. Therefore, when we look at the Trial Passages we see the same design, measurements, and angles of the Descending Passage, Ascending Passage, and the Grand Gallery. From this information we can extrapolate how the Giza power plant's development and testing took place:

- *The Grand Gallery Resonators.* Instead of testing all twenty-seven resonators in an expensive building, they may have been developed and tested in groups of two or three in the Trial Grand Gallery. This testing may have been accomplished by simulating the vibrations from the Earth and directing sound up the Trial Ascending Passage.

- *The King's Chamber Granite Complex.* The development and testing of the complex, including the Antechamber acoustic filter, may have been accomplished elsewhere and, for testing purposes, may not have needed to rely on the work being performed in the Trial Passages. Sound could have been simulated and focused through the acoustic filter to accomplish this. There are marks on the granite beams above the King's Chamber that establish their unique positions when being installed in the Great Pyramid. This is evidence that it was very important that each piece be installed in exactly the same position as when it was developed and tested.

- *The Queen's Chamber Reaction Chamber.* There are no Trial Passages that correspond to this feature. There is a slight indication of a Horizontal Passage, but this was probably cut to test the slab that bridges the Grand Gallery and the Ascending Passage. Considering the function of the Queen's Chamber, the fact that this chamber was not included in the Trial Passages is perfectly logical. It would be a waste of time to dig a long tunnel with a chamber at the end of it to fulfill a purpose that could be handled easily in an above-ground laboratory. They could have generated hydrogen without digging the Horizontal Passage and the Queen's Chamber, so why bother digging them?

Considering the investment the ancient Egyptians made in building such a structure, and its intended purpose as a power plant, it would be nearly unthinkable for them not to have fully tested the machinery that would be put to use. The remarkable similarity in the dimensions of both the passages in the Great Pyramid and the Trial Passages supports my speculation that every piece of equipment critical to the operation of the power plant was first fully developed and tested prior to its installation. The power plant theory currently is the only one that provides a logical pattern of events to explain the purpose for the Trial Passages.

With our knowledge of the progress of modern science, we can easily visualize the events that transpired after the builders successfully tested their equipment in their trial laboratory. Construction, if not already begun, would have been started on the largest pyramid ever built, and the equipment to be used inside would have been stored until it was convenient to install it permanently. Perhaps some improvements could have been made in the meantime, with respect to the efficiency of the equipment in terms of operation and longevity, for, once installed, it would have to operate without adjustment, maintenance, or any other human intervention. Thus, every dimension would had to have been cross-checked first and approved by the chief designer before the quarries were given the final go ahead to cut the stone that would house this equipment of the Giza power plant. Once the pyramid itself had been completed and the equipment installed, the workers would have packed their tools and headed home while the contracting company gave the key that operated this state-of-the-art system to its new owners.

Chapter Thirteen

SUMMARY

I f my power plant theory was based on evidence from a singular *exhibit or a few artifacts, critics and skeptics could rightly at-tribute that evidence to pure coincidence. However, I have amassed a plethora of facts and deductions based on sober con-sideration of the design of the Great Pyramid and nearly every artifact found within it that, when taken together, all support my premise that the Great Pyramid was a power plant and the King's Chamber its power center. Facilitated by the element that fuels our sun (hydrogen) and uniting the energy of the universe with that of the Earth, the ancient Egyptians converted vibrational energy into microwave energy. For the power plant to function, the designers and operators had to induce vibration in the Great Pyramid that was in tune with the harmonic resonant vibrations of the Earth. Once the pyramid was vibrating in tune with the Earth's pulse it became a coupled oscillator and could sustain the transfer of energy from the Earth with little or no feedback. The three smaller pyramids on the east side of the Great Pyramid may have been used to assist the Great Pyramid in achieving resonance, much like today we use smaller gasoline engines to start large diesel engines. So let us now turn the key on this amazing power plant to see how it operated (see Figure 69).

The Queen's Chamber, located in the center of the pyramid, and di-rectly below the King's Chamber, contains peculiarities entirely different than those observed in the King's Chamber. The Queen's Chamber's characteris-tics indicate that its specific purpose was to produce fuel, which is of para-mount importance for any power plant. Although it would be difficult to pinpoint exactly what process took place inside the Queen's Chamber, it appears a chemical reaction repeatedly took place there. The residual sub-stance the process left behind (the salts on the chamber wall) and what can

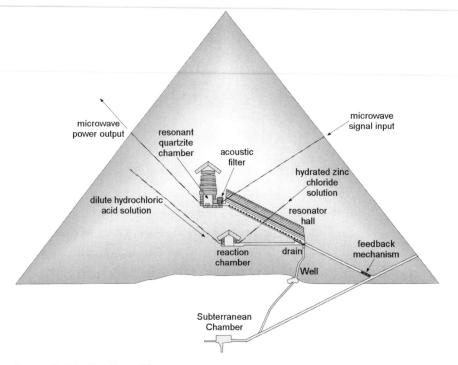

FIGURE 69. *The Giza Power Plant*

be deduced from artifacts (grapnel hook and cedarlike wood) and structural details (Gantenbrink's "door" for example) are too prominent to be ignored. They all indicate that the energy created in the King's Chamber was the result of the efficient operation of the hydrogen-generating Queen's Chamber.

The equipment that provided the priming pulses was most likely housed in the Subterranean Pit. Before or at the time the "key was turned" to start the priming pulses, a supply of chemicals was pumped into the Northern and Southern Shafts of the Queen's Chamber, filling them until contact was made between the grapnel hook and the electrodes that were sticking out of the "door." Seeping through the "lefts" in the Queen's Chamber, these chemicals combined to produce hydrogen gas, which filled the interior passageways and chambers of the pyramid. The waste from the spent chemicals flowed along the Horizontal Passage and down the Well Shaft.

Induced by priming pulses of vibration—tuned to the resonant frequency of the entire structure—the vibration of the pyramid gradually increased in amplitude and oscillated in harmony with the vibrations of the Earth. Harmonically coupled with the Earth, vibrational energy then flowed in abundance from the Earth through the pyramid and influenced a series of tuned Helmholtz-type resonators housed in the Grand Gallery, where the vibration was converted into airborne sound. By virtue of the acoustical design of the Grand Gallery, the sound was focused through the passage leading to the King's Chamber. Only frequencies in harmony with the resonant frequency of the King's Chamber were allowed to pass through an acoustic filter that was housed in the Antechamber.

The King's Chamber was the heart of the Giza power plant, an impressive power center comprised of thousands of tons of granite containing fifty-five percent silicon-quartz crystal. The chamber was designed to minimize any damping of vibration, and its dimensions created a resonant cavity that was in harmony with the incoming acoustical energy. As the granite vibrated in sympathy with the sound, it stressed the quartz in the rock and stimulated electrons to flow by what is known as the piezoelectric effect. The energy that filled the King's Chamber at that point became a combination of acoustical energy and electromagnetic energy. Both forms of energy covered a broad spectrum of harmonic frequencies, from the fundamental infrasonic frequencies of the Earth to the ultrasonic and higher electromagnetic microwave frequencies.

The hydrogen freely absorbed this energy, for the designers of the Giza power plant had made sure that the frequencies at which the King's Chamber resonated were harmonics of the frequency at which hydrogen resonates. As a result, the hydrogen atom, which consists of one proton and one electron, efficiently absorbed this energy, and its electron was "pumped" to a higher energy state.

The Northern Shaft served as a conduit, or a waveguide, and its original metal lining—which passed with extreme precision through the pyramid from the outside—served to channel a microwave signal into the King's Chamber. The microwave signal that flowed through this waveguide may have been the same signal that we know today is created by the atomic hydrogen that fills the universe and that is constantly bombarding the Earth. This microwave signal probably was reflected off the outside face of the pyramid, then was focused down the Northern Shaft. Traveling through the King's Chamber and passing through a crystal box amplifier located in its path, the input signal increased in power as it interacted with the highly energized hydrogen atoms inside the resonating box amplifier and chamber. This interaction forced the electrons back to their natural "ground state." In turn, the hydrogen atoms released a packet of energy of the same type and frequency as the input signal. This "stimulated emission" was entrained with the input signal and followed the same path.

The process built exponentially—occurring trillions of times over. What entered the chamber as a low energy signal became a collimated (parallel) beam of immense power as it was collected in a microwave receiver housed in the south wall of the King's Chamber and was then directed through the metal-lined Southern Shaft to the outside of the pyramid. This tightly collimated beam was the reason for all the science, technology, craftsmanship, and untold hours of work that went into designing, testing, and building the Giza power plant. The ancient Egyptians had a need for this energy: It was most likely used for the same reasons we would use it today—to power machines and appliances. We know from examining Egyptian stone artifacts that ancient craftspeople had to have created them using machinery and tools that needed electricity to run. However, the means by which they distributed the energy produced by the Giza power plant may have been a very different process from any we use today. Because I lack hard evidence to support any speculation about their process, I will not address that issue now, but I will offer several hypotheses in the next chapter.

Before we move into the more speculative part of the book, I would like to join architect James Hagan and other engineers and technologists in extending my utmost respect to the builders of the Great Pyramid. Though some academics may not recognize it, the precision and knowledge that went into its creation are—by modern standards—undeniable and a marvel to behold. In *Secrets of the Great Pyramid,* Peter Tompkins informed us of the opinion of professor F.A.P Barnard, of Columbia College in New York, who energetically attacked the work and ideas of Piazzi Smyth. Barnard criticized the ancient Egyptians for the "stupidly idiotic task of heaping up a pile of massive rock a million-and-one-half cubic yards in volume."[1] We can believe that the pyramid builders were primitive and that they used primitive methods of manufacturing if we choose to, but practical experience in the skills and technology that must have had a part in the creation of countless numbers of ancient artifacts in Egypt forces many people, myself included, to reject such notions. When we *know* what to look for we *cannot* ignore the evidence of advanced methods of machining! I hope this fact alone will persuade those working in the fields of archaeology and Egyptology to take another look at this material.

The evidence presented in this book, for the most part, was recorded many years ago by men of integrity who worked in the fields of archaeology and Egyptology. That much of this evidence was misunderstood only reveals the pressing need for an interdisciplinary approach to fields that have until recently been closed to nonacademics and others outside the fold of formal archaeology and Egyptology. Much of our ignorance of ancient cultures can be placed at the feet of closed-minded theorists who ignore evidence that does not fit their theories or fall within the province of their expertise. Sometimes it takes a machinist to recognize machined parts or machines! As a result, much of the evidence that supports a purpose for the Great Pyramid as anything other than a tomb has been ignored, discounted without serious consideration, or simply explained away as purely coincidental. Is it coincidence that the Great Pyramid is so huge and so precise? That the King's Chamber contains so many indications that tremendous forces disturbed it or were created within it at one time? Are the exuviae, the chocolate-colored granite, the resonating chambers with their giant granite monoliths placed above, and the unique properties of the quartz crystal present in vast quantities in the granite complex all coincidental? Can the design and physical tests of the movement of sound inside the Grand Gallery be just a happy accident? How about the series of notches along the Grand Gallery? They had to have some purpose.

We technologists can appreciate the pride the pyramid builders must have felt after they had developed their technology and the results stood majestically against the Egyptian skyline. The Great Pyramid inspired awe, which must have been enhanced by its function, enriching the lives of the people who contributed to its construction. If our society had developed a power plant that embodied the features of the Great Pyramid, there would be a renaissance in public thought regarding power-related technology and how it affects an individual's life. If the technology that can be seen inside the Great Pyramid was replicated for our benefit, there would be less concern about the future of our technological society, for a vast renewable source of energy would be available for as long as we inhabit this planet. Water and/or simple chemicals enter at one point and energy is output from another. No pollution and no waste. What could be simpler?

Well, it may not be quite that simple. The technology that was used inside the Great Pyramid may be quite simple to understand but might be difficult to execute, even for our technologically "advanced" civilization. However, if anyone is inspired to pursue the theory presented in this book, their vision may be enhanced by the knowledge that re-creating this power source would be ecologically pleasing to those who have a concern about the environmental welfare and the future of the human race. Blending science and music, the ancient Egyptians had tuned their power plant to a natural harmonic of the Earth's vibration (predominantly a function of the tidal energy induced by the gravitational effect that the moon has on the Earth). Resonating to the life force of Mother Earth, the Great Pyramid of Giza quickened and focused her pulse, and transduced it into clean, plentiful energy.

Besides obvious benefits from such a power source, we also should consider the benefits that could be gained by utilizing such a machine in geologically unstable areas of the planet. As we discussed earlier, over time there is an enormous amount of this energy built up in the Earth. Eventually the weak spots in the mantle can give way to these stresses, releasing tremendously destructive forces. If we could build a device to draw mechanical energy from seismically active regions of the planet in a controlled fashion—instead of it accumulating to the destructive level of earthquakes—we might be able to save thousands of lives and billions of dollars. We would have a device that would help stabilize the planet. So rather than being periodically shaky real estate, California might eventually become the United States' energy mecca, with a Great Pyramid drawing off the energy that is building up within the San Andreas Fault. A fanciful idea? Perhaps not.

If we assume that there are no coincidental features in the Great Pyramid, then the ancient Egyptians have proved that they were knowledgeable about the dimensions of the Earth, as well as its physical relationship to the sun and the moon. We can reasonably speculate that the knowledge of astronomy, embodied in the Great Pyramid, was not coincidental or a fanciful idea of the builders, but a necessary element in tuning their power plant to the pulse of our dynamic Earth.

We know very little about the pyramid builders and the period of time when they erected these giant monuments; yet it seems obvious that the entire civilization underwent a drastic change, one so great that the technology was destroyed with no hope of rebuilding. Hence a cloud of mystery has denied us a clear view of the nature of these people and their technological knowledge. Considering the theory presented in this book, I am compelled to envision a fantastic society that had developed a power system thousands of years ago that we can barely imagine today. This society takes shape as we ask the logical question, "How was the energy transmitted? How was it used?" These questions cannot be fully answered by examining the artifacts left behind. However, these artifacts can stimulate our imaginations further; then we are left to speculate on the causes for the demise of the great and intelligent civilization that built the Giza power plant. This speculation is the subject of the remainder of this book.

Chapter Fourteen

A GLIMPSE INTO THE PAST

ur lives are dependent on the switch. How many times do we use the switch in a day? Ever count them? Have you ever followed the wire back to the source, in your mind, and paused in wonderment at the true power you have at your fingertips? How many miles of electrical cable will your mind travel along before it reaches the 500-megawatt turbine generators at the power station? Now think about what life would be like without electricity. Actually, we do not have to go too far back in time to relive that scenario. Every device that uses electricity has been developed within the past one hundred years. We are now so dependent on electricity and the switch that it would be inconceivable to be without them. Some of us will remember when there were no electric lights in our homes, and gaslight and candles provided illumination. To others, such an existence will be beyond comprehension. The electricity that feeds the homes of developed countries is synonymous with shelter and has become as basic a staple as food and clothing. How did we come so far so fast?

Modern electrical power distribution technology is largely the fruit of the labors of two men—Thomas Edison and Nikola Tesla. Compared with Edison, Tesla is relatively unknown, yet he invented the alternating electric current generation and distribution system that supplanted Edison's direct current technology and that is the system currently in use today. Tesla also had a vision of delivering electricity to the world that was revolutionary and unique. If his research had come to fruition, the technological landscape would be entirely different than it is today. Power lines and the insulated towers that carry them over thousands of country and city miles would not distract our view. Tesla believed that by using the electrical potential of the Earth, it would be possible to transmit electricity through the Earth and the atmosphere without using wires. With suitable receiving devices, the electricity could be used in remote parts of the planet. Along with the transmission of electricity, Tesla proposed a system of global communication, following an inspired realization that, to electricity, the Earth was nothing more than a small, round metal ball. In a letter to *Electrical World and Engineer* magazine, March 5, 1904, Tesla wrote:

> *When the great truth, accidentally revealed and experimentally confirmed is fully recognized, that this planet, with all its appalling immensity, is to electric currents virtually no more than a small metal ball and that by this fact many possibilities, each baffling imagination and of incalculable consequence, are rendered absolutely sure of accomplishment; when the first plant is inaugurated and it is shown that a telegraphic message, almost as secret and non-interferable as a thought, can be transmitted to any terrestrial distance, the sound of the human voice, with all its intonations and inflections, faithfully and instantly reproduced at any other point plying light, heat or motive power, anywhere—on sea, or land, or high in the air—humanity will be like an ant heap stirred up with a stick: See the excitement coming.*[1]

With $150,000 in financial support from J. Pierpont Morgan and other backers, Tesla built a radio transmission tower at Wardenclyffe, Long Island, that promised—along with other less widely popular benefits—to provide communication to people in the far corners of the world who needed no more than a handheld receiver to utilize it.

In 1900, Italian scientist Guglielmo Marconi successfully transmitted the letter "S" from Cornwall, England, to Newfoundland and precluded Tesla's dream of commercial success for transatlantic communication. Because Marconi's equipment was less costly than Tesla's Wardenclyffe tower facility, J. P. Morgan withdrew his support. Moreover, Morgan was not impressed with Tesla's pleas for continuing the research on the wireless transmission of electrical power. Perhaps he and other investors withdrew their support because they were already reaping financial returns from those power systems both in place and under development. After all, it would not have been possible to put a meter on Tesla's technology—so any investor could not charge for the electricity!

Without the support of Morgan, Tesla's other sources for finance dried up and he became depressed and infirm. He was forced to leave his opulent apartment at the Waldorf Astoria hotel; and his partially manifested dream at Wardenclyffe was torn down and sold for scrap to pay off his debt. As Marconi's stature and fortune rose, Tesla's decline drained his vitality; he passed from this life in a New York hotel room in 1943, leaving a legacy that, even today, still inspires and feeds the intellect of researchers all over the world.

It is distinctly clear that tenuous timing and fickle social circumstances surrounding key inventors and their inventions have shaped our techno-logical landscape. The greatest influence on whether an invention comes to fruition or not is its investors' desire for profit. If Tesla had succeeded before Marconi in achieving transatlantic transmission, would our power delivery systems look like they do today? Perhaps they would if the motivating force behind their construction had been only profit, a profit realized through the metered flow of electricity. It would have been next to impossible to con-vince investors to give for free what was already reaping handsome rewards. Would such restraints on innovation occur today? Perhaps, but the technol-ogy for metering the use of wireless electricity could be accomplished in the same way telephone companies charge for cellular calls. The voice you hear on your cell phone is actually wireless transmitted energy—albeit very little—that is sold to you at a metered rate.

Other, more recent proposals for alternative power delivery also have not been realized on a large scale. The inhabitants of Reykjavik, Iceland, heat their homes and power their processing plants with a natural resource—geothermal energy. Icelanders enjoy year-round benefits in their geother-mal heated swimming pools. They have such an abundance of geothermal power that at one time they proposed selling their surplus energy to other countries. Because it would be impractical to build steam pipes across the ocean (not that they even thought of such a folly), an idea was put forward that combined the technology of geothermal energy with some popular sug-gestions for harnessing solar energy by geosynchronous satellites and trans-mitting the energy to Earth via microwave beam.

FIGURE 70. *Egyptian Relay Satellite*

If a satellite could harness solar energy, convert it into electromagnetic energy, and transmit this energy through space to be collected on the Earth, then Earth-based power plants could convert their energy to microwave energy and transmit it into space. A collimated microwave beam could be directed into space to a passive microwave reflector satellite and reflect to a distant point on the Earth (see Figure 70). A microwave beam can pass through clouds and rain with very little attenuation, or loss of energy. A ground-based antenna could then convert it to usable electrical power. It is even possible that a series of satellites could split the energy and deliver it to several points around the globe. Such an energy distribution method is technically feasible, but, like many other proposals for technological innovation, the funding necessary to bring it into physical manifestation is not always immediately forthcoming.

The point of this discussion is that there are many viable energy systems, but those chosen for use are often the ones that make *economic,* not technological, sense. We must keep in mind, therefore, that what makes sense to us may not have made sense to past cultures. When we try to envision past energy systems we have many layers of cultural blinders to see through. As we search through the remnants of ancient Egypt looking for the power plants that provided energy to

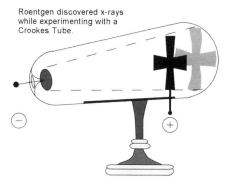

Roentgen discovered x-rays while experimenting with a Crookes Tube.

Accelerated electrons fly past the anode and form a shadow on the wall of the tube.

FIGURE 71. *Crookes Tube*

the machine tools that accurately shaped the granite blocks on the Giza Plateau or the granite boxes in the rock tunnels at Saqqara, we cannot assume that their power plants looked like ours or that the infrastructure supporting the distribution of energy was the same. Considering the extremely tenuous circumstances by which inventions are developed, promoted, and utilized, it would be very surprising to find an ancient artifact, or evidence of an artifact, that is identical to one we use or have used in the recent past.

This is why I was flabbergasted and stunned when, while looking through a chemistry book one day, I came across an illustration of a Crookes tube (see Figure 71). I had seen such an electrical device before—in photographs taken of an Egyptian temple! The wall carvings at Dendera in the lower crypt in the temple of Hathor contain an image that looks similar to a Crookes tube (see Figure 72).

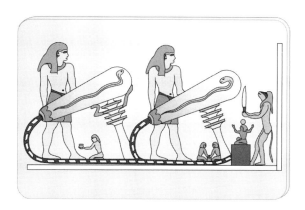

FIGURE 72. *Wall Carving at Dendera*

Then, while I was working on the final stages of this book, I came across another reference to the wall carving at Dendera and its graphical representation of a Crookes tube in Brad Steiger's book *Worlds Before Our Own*. As I had read this book in 1978, the information probably receded into my subconscious mind and, without a conscious link to this book, resurfaced only after I had actually seen the illustrations many years later. The question of how the ancient Egyptians were able to illuminate the passageways and chambers in their pyramids and tombs has puzzled many people; the walls and ceilings of the tunnels and chambers are not marked with the smut that would accumulate in the use of blazing torches. While pondering this phenomenon, Steiger referred to the research of Joey R. Jochmans who presented an analysis of the wall carvings at Dendera:

> When the [Crookes] tube is in operation, the ray originates where the cathode electrical wire enters the tube to the opposite end. In the temple picture, the electron beam is represented as an outstretched serpent. The tail of the serpent begins where a cable from the energy box enters the tube, and the serpent's head touches the opposite end. In Egyptian art, the serpent was the symbol of divine energy.
>
> ... The Temple picture shows one tube, on the extreme left of the picture, to be operating under normal conditions. But with the second tube, situated closest to the energy box on the right, an interesting experiment has been portrayed. Michael R. Freeman, an electric and electromagnetic engineer, believes that the solar disc on Horus' head is a Van de Graaff generator, an apparatus which collects static electricity. A baboon is portrayed holding a metal knife between the Van de Graaff–solar disc and the second tube. Under actual conditions, the static charge built up on the knife from the generator would cause the electron beam inside the Crookes tube to be diverted from the normal path, because the negative knife and negative beam would repel each other. In the Temple picture, the serpent's head in the second tube is turned away from the end of the tube, repulsed by the knife in the baboon's hand.[2]

Steiger presented another analysis by a professional engineer, who saw the wall carvings at Dendera as an accurate illustration of an electrical device—one which would not be out of place in a modern electrical blueprint file. "In regards to the ancient Egyptian electron tubes, electromagnetic engineer Professor S.R. Harris identified a box-and-braided cable in the picture as 'virtually an exact copy of engineering illustration used today for representing a bundle of conducting wires.' The cable runs from the box the full length of the floor and terminates at both the ends and at the bases of two peculiar objects resting on two pillars. Professor Harris is said to have identified these representations as high voltage insulators."[3]

While technologists compare the image in this wall carving with what they know of science, Egyptologists interpret these images with what they know of ancient Egyptian symbolism. Unorthodox Egyptologist, John Anthony West, is convinced that Egyptian symbolism fully explains these images without having to invoke higher scientific knowledge. According to West, the carvings represent a manifestation of consciousness, with the serpent born aloft by the lotus—symbols representing a cosmogonical principle underlying all creation. As these are only illustrations and not the object they actually represent, different interpretations of what they actually mean may always exist. They are interesting graphics, in light of what is presented in this book, but not an essential part of my theory.

Even though I have solid evidence to support my Giza power plant theory, I am still haunted with questions that challenge this radical notion that the ancient Egyptians had electricity. "What did they do with the energy they produced? Where are the machines?" I often hear these questions when discussing my power plant theory with others. Of course, when we consider a civilization that is blessed with abundant electrical energy, we immediately think of refrigerators, washers, dryers, and other appliances that have become so necessary to our own comfortable existence. Not being able to identify any museum with a prehistoric toaster oven on display, I generally have responded that once a society develops a power system, it may or may not go on to invent the kinds of devices to use that power. But with the Crookes tube parallel, we have a hint that at least one of the devices the ancient Egyptians had developed had inspired an artisan to carve its likeness in stone.

The Great Pyramid a power plant. The carving at Dendera an electrical device? Igneous rock at numerous locations throughout Egypt that shows signs of having been precision machined. The evidence is there that the ancient Egyptian civilization was a lot different, and more advanced, than historians have led us to believe. So what did the Egyptians do with the energy they produced? From the exit point of the Southern Shaft of the Great Pyramid, the energy that flowed fulfilled a need for the civilization that invested in its construction. How was this energy delivered to those who used it? Our route to an answer takes us full circle, and we find ourselves starting at the Giza power plant itself.

As evidenced by their ability to lift huge weights, both Edward Leedskalnin and the ancient Egyptians were utilizing technology that we do not possess. Their ability to use gravity against itself and make large masses weightless forecasts the development of new technology which may include vehicles that use very little energy and that, conceivably, could gently break through the Earth's atmosphere, hover indefinitely at any point in space, and then safely return to Earth. A society that is not bound by the effects of gravity is a society that is finally unchained from the primitive wheel (which Egyptologists note the Egyptians did not use) and the wasteful, albeit sophisticated, use of fire (such as a jet engine). Such a society would no longer need to burn up the planet. Dare we speculate that the energy that exited the Southern Shaft of the Great Pyramid may have been directed to a point in space where a satellite collected the energy and beamed it to a distant point on Earth? Perhaps that point was within the borders of the mythical Atlantis. Perhaps the use for this energy was even more fantastic than we have ever before dreamed.

Robert Bauval is a man who dares to venture into the land of the fantastic. He has an incredible passion for his work and for the theory he proposed, with Adrian Gilbert, in *The Orion Mystery.* I first met Bauval in February 1995 when he, along with Graham Hancock and Netherlands Television producer Roel Oostra, knocked on my door at the Movenpick hotel near Giza. They were making a courtesy call to inform me that the following morning, after breakfast, they planned to do some filming with me in the King's Chamber inside the Great Pyramid. I had just completed a long journey and was tired, so after their visit I lay down to rest and replayed the events that had brought me to Egypt.

I had concluded many years ago that the ancient Egyptians left nothing, in their design and construction, to chance. When I saw *The Orion Mystery* in Paul and Ardith Keller's bookstore in Camby, Indiana, and leafed through it for a few minutes, I knew I had to buy it. I was holding in my hands a new and revolutionary theory on the pyramids of Egypt.

The Orion Mystery described the placement of the three major pyramids on the Giza Plateau to be analogous to the belt of the constellation Orion. The Great Pyramid and the Second Pyramid (Khafra's) are in alignment and close in size, and the Third Pyramid (Menkaura's) is smaller and offset from the others, mirroring the spatial relationship of the three stars of Orion's belt. With this revelation, the work of astronomer Virginia Trimble became more valuable. Trimble calculated that the Southern Shaft leading from the King's Chamber, with its angle of 44.5 degrees, was in alignment with Orion's belt. The Pyramid Texts associate Orion with the god of the afterlife, Osiris. Trimble surmised that after death, the king's soul traveled up the Southern Shaft to Orion and the king was reborn as Osiris. As Bauval describes Trimble's and his revelation in a documentary, he was indeed inspired with this relationship, and he carried this inspiration for over twelve years before finally writing about it in his best-selling book in 1994. It was the autumn of 1994 that I read the book and sent a letter to Bauval along with my article "Advanced Machining in Ancient Egypt."

Bauval introduced me to Graham Hancock, who was writing *Fingerprints of the Gods* at the time, and subsequently to Roel Oostra, who invited me to Egypt to participate in a documentary. I believed in Bauval's theory—with the exception that he still maintained the tomb theory. Like him, however, I do not believe the Southern Shaft alignment and the placement of the three Giza pyramids happened by chance. For me, any references in the Pyramid Texts to the king's soul traveling through the Southern Shaft and rising to Orion are metaphors. I am, therefore, in agreement with Bauval and Gilbert who related the process of interpreting these hieroglyphics as analogous to reading a word processing file in a computer:

> *Anyone who has worked with a computer knows that calling up a file using a word-processing program not compatible with the one being used, means a garbled version of the text appearing on the screen.*
> *This is more or less what happened (and in many ways is still*

happening) with the Pyramid Texts and the pyramids of Egypt. We
believe that the wrong program for reading them has been used. We
are not talking of the translation from the hieroglyphic language to
modern languages; we have the utmost faith in the work of Faulkner
and others like him. We are referring specifically to the interpreta-
tion put on these texts by Egyptologists. We believe that the proper
program or decoder exists and needs to be understood before we can
properly decode the Pyramid Texts and extract their real, esoteric
meaning.[4]

Their idea prompted me to consider my own theory in another way. Given the astronomical association and significance of the Southern Shaft, is it possible that the energy produced in the King's Chamber, and which then exited the Southern Shaft, *is* referred to in the Pyramid Texts? Perhaps I had only to apply the correct algorithm to interpret it. I read on in *The Orion Mystery*, noting a recurring metaphor from the Pyramid Texts:

'*. . . The king is a Star . . .*' [PT 1583]
'*The king is a Star which illuminates the sky . . .*' [PT 362, 1455]
'*. . . The king is a Star brilliant and far-travelling . . . the king*
appears as a star . . .' [PT 262]
'*Lo, the king arises as this star which is on the underside of the*
sky . . .' [PT 347]
There can be little doubt that the Pyramid Texts make a clear
statement that the dead kings become stars, especially seen in the lower
eastern sky. They also tell us that it is the souls of departed kings
which become stars:
'*be a soul as a living star . . .*' [PT 904]
'*I am a soul . . . I (am) a star of gold . . .*' [PT 886–9]
'*O king, you are this Great Star, the companion of Orion, who*
traverses the sky with Orion, who navigates the Duat with Osiris . . .'
[PT 882][5]

What an incredible description of events related to the pyramids' link to the stars. Suddenly I saw a new meaning for this star. Imagine if we had put a vehicle in space, for whatever reason, and were beaming energy to it—for the vehicle's own use or to be returned to some location on Earth. Would that vehicle not appear as a bright star in the night sky? Assuming that the energy beam would have some divergence (it would grow in size) the farther it traveled from its source, then the larger the microwave receiver—the "star"—would have to be. On a clear night, we can see a small satellite as it orbits the Earth. What an incredible view we would have of such a receiver dish (refer to Figure 70). Looked at from an angle, would it appear as an eye in the sky?

And more fantastic still, what if the energy were being used to provide power to a space ship? The microwave energy that was projected from the Southern Shaft to Orion's belt stars may have been delivering more than Khufu's soul to the heavens. The energy, which Robert Bauval describes as Khufu's soul traveling to Orion, may have been his actual person along with an entourage!

If this seems too fantastic a speculation, take a moment to consider how future civilizations might interpret our own account of the Sojourner expedition to Mars. Five thousand years from now they probably would not be able to make sense of it because the media that contains the information would have degraded by then. We are in the same position with respect to the ancient Egyptians as our far future descendants will be to us. The Egyptians left multiple records of this "star" carved in stone. Have they been interpreted correctly? Is the interpretation of the hieroglyph of "king" correct?

In the English language there are different meanings for the word "power." It can mean strength, ability, or authority, as in a leader (king). It can also refer to the energy contained in a battery or delivered to your home through a wire. Is it possible that the Pyramid Texts refer to an environment associated with the production of power that was projected into space? We may never be able to answer this question, but it is one that is worth asking and thinking about.

Our speculations on the relationship between the Giza power plant theory and the interpretation of the Pyramid Texts could go on endlessly. For now, they may seem to be more in the realm of science fiction, but we must keep in mind that some of the technology that exists today, and that has been described in this book, was originally described in works of science fiction many years ago. Space travel, global communication satellites, and handheld remote communication devices were technologies that inspired generations of scientists and engineers during their formative years as they devoured the imaginative yarns spun by science fiction writers.

And we must push our speculations even harder. The interpretation of ancient Egyptian hieroglyphs by Egyptologists presents us with a view of a civilization that placed tremendous emphasis on the afterlife, with the kings and pharaohs being preoccupied with life after death. Since Neanderthal tribes first started to bury their dead, humans have held the belief that there is more to our existence than what our physical senses can detect. Many of us require three-dimensional definitions for the things that surround us, but all of natural phenomena cannot be described in these terms—if they could there would be no more mystery or research.

I am compelled, therefore, to touch on a subject that I have been advised to avoid because it brings an element that is not three-dimensional to my work. Every three-dimensional object we use and enjoy today had its beginnings with some form of inspiration or speculative thought. This inspiration may be characterized differently by different people, but common to all is the fact that creativity transcends what we know as physical "reality." If this book is going to be complete and honest, I cannot ignore the information that came into my hands recently in the form of a small blue paperback entitled *Edgar Cayce on Atlantis*.

My wife, Jeanne, has this book in her library, and she compelled me to read what Edgar Cayce had to say about Atlantis because of the flurry of activity he has generated in Egypt. Edgar Cayce, also known as the sleeping prophet, has influenced powerful people in all walks of life. His research foundation in Virginia Beach, Virginia, is the site of yearly conferences at which both scholars and avant-garde researchers have met to discuss and debate issues pertaining to the long-sought-after Atlantean Hall of Records that Cayce predicted would be found near the Great Sphinx.

The jostling for recognition in the historical record is no different today than it was in 1922 when archaeologist Howard Carter discovered the tomb of King Tutankhamen. Recent discoveries made public by Boris Said and Tom Danley spring from a Cayce follower's lifelong desire to find evidence supporting one of Cayce's prophecies—that there is an Atlantean Hall of Records located near the Sphinx. The follower's name is Joseph Schor, and he funded Said and Danley's sonic tests inside the Great Pyramid and around the Sphinx, as well as the exploration of a deep shaft that was discovered close to the causeway nearby. Dr. Zahi Hawass, the director of the Giza Plateau, facilitated their activities, perhaps hoping for an increase in tourism that would follow such a discovery. Hawass has shown support for the discovery of Cayce's Hall of Records and teased before the camera in a tunnel near the Sphinx with news of what was described as a new chamber, which would be opened on live television. The promotional video was not meant to be released, but it was and Hawass's excited promo was revealed to the public prematurely.

Some readers may think Cayce's psychic speculations have no place in a serious book on Egypt. But in light of the fact that the Cayce foundation funded Egyptologist Mark Lehner in his studies in Egypt in the 1970s—though Lehner ultimately moved away from the foundation and he became a staunch supporter of the orthodox view—I feel justified in discussing Cayce's work and, perhaps, putting it into a different perspective. I should add, though, that I do not include this discussion to endorse Cayce's readings, nor have Cayce's readings held up as support for my theory. This is included because Cayce dealt a great deal with ancient Egypt, and his description of ancient technology while in his altered state cannot be ignored.

Cayce's son, Edgar Evans Cayce, does an admirable job explaining his father's psychic readings, which were produced while Cayce was in trance. A series of readings by Cayce—known collectively as reading number 440—with an engineer, referred to in the reading as "the entity," describes technologies used in Atlantis and Egypt that have an amazing similarity to the technology described in the Giza power plant theory. These readings described the ancient Atlantean power plants, which on the surface seem far removed from the Egyptian pyramids; however, an interpretation of the readings becomes more meaningful when we compare (what Cayce described as) the "firestone" with granite, out of which the King's Chamber, the power center in the Giza power plant, is constructed:

About the firestone—the entity's activities then made such applications as dealt both with the constructive as well as destructive forces in that period. It would be well that there be given something of a description of this so that it may be understood better by the entity in the present.

In the center of a building which would today be said to be lined with nonconductive stone—something akin to asbestos, with . . . other nonconductors such as are now being manufactured in England under a name which is well known to many of those who deal in such things. The building above the stone was oval; or a dome, wherein there could be . . . a portion for rolling back, so that the activity of the stars—the concentration of energies that emanate from bodies that are on fire themselves, along with elements that are found and not found in the earth's atmosphere.

The concentration through the prisms of glass (as would be called in the present) was in such manner that it acted upon the instruments which were connected with the various modes of travel through induction methods which made much the [same] character of control as would in the present day be termed remote control through radio vibrations or directions; though the kind of force impelled from the stone acted upon the motivating forces in the crafts themselves.

The building was constructed so that when the dome was rolled back there might be little or no hindrance in the direct application of power to various crafts that were to be impelled through space— whether within the radius of vision or whether directed under water or under other elements, or through other elements.

The preparation of this stone was solely in the hands of the initiates at the time; and the entity was among those who directed the influences of the radiation which arose in the form of rays that were invisible to the eye but acted upon the stones themselves as set in the motivating forces—whether the aircraft were lifted by the gases of the period; or whether for guiding the more-of-pleasure vehicles that might pass along close to the earth, or crafts on the water or under the water.

These, then, were impelled by the concentration of rays from the stone which was centered in the middle of the power station, or powerhouse (as would be the term in the present).

In the active forces of these, the entity brought destructive forces by setting up—in various portions of the land—the kind that was to act in producing powers for the various forms of the people's activities in the cities, the towns and the countries surrounding same. These, not intentionally, were tuned too high; and brought the second period of destructive forces to the people in the land—and broke up the land into those isles which later became the scene of further destructive forces in the land. (440–5; Dec. 20, 1933)[6]

Though Edgar Evans Cayce interprets the reading to mean that when the Atlantean power house was in operation the "firestone" was on fire, or influenced by heat, another more accurate interpretation can be made. The firestone to which Cayce refers is actually an accurate description of igneous rock—or granite (igneous, in this sense, means "produced by or resulting from the action of fire").[7] When I read the word "firestone" in reference to the Atlantean power house, I immediately thought of the King's Chamber's thousands of tons of granite that were prepared by those who would have possessed a higher knowledge of the sciences and arts. The crystal Cayce refers to, however, is contained within the granite itself in the Great Pyramid, in the form of quartz crystals—though there very well may have been an additional solid crystal that was cut and polished to amplify the microwave input signal. The application of Tesla technology using wireless transmission of electricity also is suggested in this passage, although Tesla's Wardencliffe tower research came thirty-three years before Cayce's reading, and one could argue that Cayce may have heard of this research and could have been influenced by it during the reading.

What is interesting about Cayce's reading is his reference to the energy of the stars and the use of these energies in conjunction with the energies he discussed in a reading in 1930: "in city of Peos in Atlantis—among people who gained understanding of application of nightside of life or negative influences in the earth's spheres, of those who gave much understanding to the manner of sound, voice and picture and such to peoples of that period." (2856–1; June 7, 1930)[8]

As I said, I have not been a follower of Cayce. During the early days of the research for this book, I was ignorant of what he had written on the subject. It is only recently that I have maintained a mild interest in what has been reported by him with respect to advanced cultures in prehistory. His description of the power system of the Atlanteans, though, is remarkably similar to what is found within the Great Pyramid. The only thing that does not seem to fit is the shape. Cayce's power plant has a domed structure with the awesome firestone located in the center of the structure. Perhaps the Great Pyramid originally had such a structure. Consider, also, Cayce's statement about the conversion of energy through the firestone. He says the energies were tuned too high and caused widespread destruction. This description certainly gives us pause to consider what changes would take place or what forces could be unwittingly unleashed within the Earth if such a power system was replicated. What unknown changes might we be setting in motion? Would we be faced with annihilation? Would we find ourselves on the threshold of our past?

In this chapter we have asked a lot of questions, only sometimes finding an answer. But our speculations are worthwhile nonetheless. All the evidence found clearly shows that the ancient Egyptians were technologically advanced and that the Great Pyramid provides us with a guidepost, a geodetic marker, that leads us not only to reconsider what we know about past civilizations but also to wonder where our own technology may ultimately lead us. If Cayce's reading has any basis in truth, we are faced with the realization that technology—no matter what the technology is or in what epoch it is developed—can have both positive and negative consequences for the civilizations that pursue its benefits. That is the subject we turn to next.

Chapter Fifteen

LESSONS FROM THE PAST

hile in Egypt, I was surprised to learn from an Egyptologist at the Citadel that the outer casing stones of the Great Pyramid were not plundered to construct the mosques of Cairo, as it has been reported, but were eroded by sandstorms over time. I questioned him as to why the Second Pyramid still had its casing stones intact at the top if this was true, but he didn't have an answer. My impression, while walking around the nine-pyramid complex, was that the stones had actually been shaken loose by some event that caused a buildup of vibration to the point that the pyramids became runaway vibrators. The pyramid at Meidum, with its outer casing stripped and piled high around its base, would certainly fit this description of events. The stones on this pyramid were stripped off through the action of forces other than the picks of mosque builders. So which theory is correct? Perhaps the question really is, Which theory is more likely? The evidence can be interpreted many ways. When I look far into the past, I can discern the existence of sciences of which there are no surviving records. They have either been destroyed or complete records never existed. Did science account for the amazing artifacts I had seen throughout Egypt? And did science also explain, at least in part, this culture's demise? When I looked for an event in Egyptian history that would explain the destruction of this culture and at the same time explain the erosion of the pyramids, I found a clue in the 1985 discovery of volcanic ash twenty feet underground in the Nile delta. This ash was found to be identical to that from an enormous eruption that occurred approximately 3,500 years ago on the Greek island of Santorini. The eruption is estimated to have been 22,000 times more destructive than the atomic bomb that was dropped on Hiroshima. Here was a partial answer. However, it was becoming clearer to me that another reason for the

destruction of a civilization can be related to its use and/or abuse of the technology it develops.

It is reasonable to assume that if we were to destroy ourselves through nuclear holocaust, the geological and biological record would bear witness to it and reveal that knowledge to future archaeologists as they became more advanced in their science. At the same time, some of our civil engineering projects might survive, and the occasional archaeological anomaly might turn up to promote some thought in that direction. Perhaps a granite surface plate would be found, and someone would puzzle over the positioning of the holes drilled into it. Who knows, maybe some future "primitive" tribe would see some significance to this plate and make a ritual object out of it.

When we predict the events that might transpire in the future, we can only draw upon our own experiences and how we interpret our ancestors' behavior. Therefore, many of our assumptions may not have any relationship to the truth at all. Our understanding of the past, however, is supported not only by our understanding of human behavior, but also by what the earlier humans have left behind. From archaeological remains, modern humans are constantly evaluating the progress of human development on this planet.

Because of the ample evidence that the ancient Egyptians accomplished enormous engineering feats, especially the mighty pyramids, we can begin to fashion a different picture of their civilization. This picture becomes even clearer when we evaluate the methods they must have used to create some of their stonework: machining methods largely unknown until just before the beginning of the twentieth century, and one that has only recently joined the family of machine tools—ultrasonics. All this evidence commands that we take a closer look at this society we have long thought to be primitive.

And yet there is so much that is missing! So much, in fact, that we have to ask ourselves if it is possible that they had the knowledge and scientific capability to destroy themselves, just as we have today? And did this destruction actually come to pass?

World history records the rise and fall of many empires and, with their endings, the vast destruction of property. It occurred to me that in the case of the earlier Egyptians, the cause of their demise was perhaps a little more catastrophic than any other historically recorded downfall or disaster. How catastrophic? If we look closely at the following evidence, perhaps we will begin to understand what such an event could have been.

We now know that the ancient Eygptians had a higher level of science and technology than what has been previously supposed. We also know that the development of technology and machines that harness and control the forces of nature holds negative—perhaps even catastrophic—consequences to those who do not control or use them wisely. When we consider the development of the atom bomb, nuclear power plants, even the automobile, it is clear that the potential for harm is present with every machine in existence.

But how far did the ancients develop their technology? Some researchers have suggested that civilizations in antiquity had actually developed and used atomic power! On the surface it may sound rather incredible, but then there is so much that has not survived the ancient pyramid builders, it would not be out of order to see what these researchers have to say about it.

Brad Steiger presented a forceful argument that in prehistory nuclear explosions had affected several areas of the Earth. He cited the discovery of fused green glass in deep stratas of the Earth, and in Gabon, Africa, the Euphrates Valley, the Sahara Desert, the Gobi Desert, the Mojave Desert, and Iraq. These vast wastelands of melted sand can be compared with White Sands, New Mexico, where the sands were fused as a result of nuclear bomb testing. Steiger wrote, "Perhaps the most potentially mind-boggling evidence of an advanced prehistoric technology that might have blown its parent-culture away is to be found in those sites which ostensibly bear mute evidence of pre-Genesis nuclear reactions. . . . At the same time, scientists have found a number of uranium deposits that appear to have been mined or depleted in antiquity."[1]

The Sahara Desert was at one time fertile, not the arid wasteland it is today. Geological and archaeological evidence shows that this 3.3-million-square-mile tract of land once received ample rainfall, and the rock paintings found in the Tassili N'Ajjer Mountains show that antelopes, elephants, and many other types of animals once occupied the previously lush area. For unknown reasons, the rainfall pattern shifted, and the subsequent imbalance between rainfall and the rate of evaporation (the rain was dried up by the sun faster than it fell from the clouds) turned the area into a desert.

Since vast regions of the globe still remain unexplored, it is impossible to say how many glassy areas there are just in the Sahara. If we are looking for areas where tremendous heat influenced terrestrial characteristics—like the heat that could be produced by nuclear forces—we have to only look in the previously mentioned deserts. Although these may not in themselves "prove" that prehistoric nuclear war had created them, there are many people who believe this was the case.

If our world were affected by a cataclysmic event—such as a polar displacement, comet strike, or self-inflicted nuclear war—after 10,000 years future generations would have few clues about the level of sophistication we had achieved. It would be fair to say that many of our artifacts would be misinterpreted and misunderstood. What would remain of the "concrete jungles" we call cities? Would they reveal to future archaeologists the full scope of our technological achievements? Future civilizations would be busy developing their own technology. Their development might be along a completely different path than ours has followed, and in its early stages it would not be as advanced. At what stage in their evolution would future archaeologists recognize a computer chip for what it really is?

The artifacts that survived thousands of years following the demise of a highly developed civilization would, in large part, depend on the level of technology that had been achieved. After the ravages of time, many of our artifacts will have crumbled into dust. If future archaeologists are able to analyze and interpret surviving artifacts correctly, some of our plastics and high-tech exotic alloys may provide clues that form a rather rough sketch of the life we now enjoy. It is safe to say, though, that any high-tech artifacts under study by a future generation will be misinterpreted until the technology needed to correctly identify them has been redeveloped.

But clues to what happened to us could be discerned from sources other than humanmade objects. Nature would retain the imprint of a nuclear holocaust. For example, the release of neutrons would sharply increase the amount of carbon 14 in the atmosphere, and it would show up in biological remains, like wood, bone, and other organic material from that period of time. C_{14} is created when the reaction of cosmic rays with the ionosphere precipitates neutrons through the atmosphere. These neutrons react with nitrogen 14, creating C_{14}. Immediately upon its creation, C_{14} starts to decay. Originally it was determined to have a half-life of approximately 5,568 years. (The half-life of radiocarbon was redefined from 5570 ± 30 years to 5730 ± 40 years in 1962.) Organic material takes in C_{14} at a constant rate, and, knowing what the level of C_{14} in an object was before it died, scientists can measure the amount left in it and calculate its age. Apart from normal variations, C_{14} stays at a constant level in the Earth's atmosphere. However, modern nuclear activities have increased the level of C_{14} in the atmosphere, and subsequently in everything that lives and breathes.

When Willard F. Libby first discovered radiocarbon dating in 1947, archaeologists, and especially Egyptologists, ignored it. They questioned its reliability, as it did not coincide with the "known" historical dates of the artifacts being tested. David Wilson, author of *The New Archaeology*, wrote, "Some archaeologists refused to accept radiocarbon dating. The attitude of the majority, probably, in the early days of the new technique was summed up by Professor Jo Brew, Director of the Peabody Museum at Harvard. 'If a C_{14} date supports our theories, we put it in the main text. If it does not entirely contradict them, we put it in a footnote. And if it is completely out-of-date we just drop it.' "[2]

The radiocarbon time scale contains other uncertainties, and errors as great as 2,000 to 5,000 years may occur. Contamination of the artifact may be caused by percolating groundwater, incorporation of older or younger carbon, and contamination in the field or laboratory. Willard Libby[3] addressed the problem of contamination, and the ability to distinguish between the chemistries of life and death (the chemistries of death being the contamination). Washing techniques were then developed to separate the two.

Egyptologists have generally agreed on the dates that had been established for the time of the pharaohs. Consequently, when radiocarbon dating came back with results showing artifacts to be between two hundred and five hundred years younger than their established historical dates, the experts were not impressed. In other words, articles with a "known" date of 5,000 years were tested and, according to radiocarbon dating, were found to be only 4,500 years old. For instance, some of the wood that was found in King Tutankhamen's tomb, historically dated at around 1350 B.C., gave a C_{14} reading of 1050 B.C.

The further back into history the C_{14} researchers went, the larger the discrepancies became. The original assumption on which C_{14} dating was based was that its level in the atmosphere is the same at all times. Egyptologists and the carbon-dating scientists were, therefore, in contradiction with each other. The Egyptologists and the archaeologists would not budge, and so the scientists were forced to reevaluate their findings, and they searched for an accurate method of calibrating C_{14} to validate its usefulness as an archaeological tool. Until that was accomplished, doubt prevailed.

The answer came in the form of tree-ring dating, and the tree that eventually provided the means to accomplish this accurate C_{14} dating was the bristlecone pine, indigenous to the southwestern United States. As the oldest living tree on Earth, the bristlecone pine enabled scientists to develop the chronology to calibrate carbon dating and "adjust the clock." The results are noteworthy. It turned out that the Egyptologists and the archaeologists were correct in their dates and the original C_{14} results were in error. In some cases, for distant dates, the error was as much as eight hundred years. But this finding had more than one interpretation: The Egyptologists may be correct in their historical timeline; or there may have been an unexplained "infusion" of C_{14} into the atmosphere at some prehistoric time. David Wilson summed up the argument this way: "If present day measurements of the radiocarbon remaining in objects which died in, say, 2,500 B.C. give a date of 2,000 B.C., then there is 'too much' carbon 14 left undecayed—perhaps it is that there was 'too much' carbon 14 in the object originally in 2,500 B.C. This is now generally accepted as being the case, but that still leaves the question open as to why there was more carbon 14 in the atmosphere and biosphere."[4] The question is still open, although scientists have speculated that if the latter scenario is true—there was more C_{14} in the ancient atmosphere than they would expect—the answer might be that variations in the Earth's magnetic field allowed increased amounts of cosmic rays to react with the ionosphere.

When carbon dating was first being developed, organic samples were collected for testing from around the world. The stipulation on the kind of samples that were collected was that they had died and ceased to draw carbon in from the atmosphere before the advent of our industrial age, and especially before nuclear testing had been carried out. The explosion of nuclear devices releases neutrons that would result in an elevation of C_{14} in the atmosphere. Tree-ring dating had revealed that there was an elevation of C_{14} in the atmosphere and in artifacts older than 1,000 B.C., which had thrown off the atomic clock. Around 8,000 B.C. to 10,000 B.C., the level of C_{14} started to fall back to "normal."

What we are forced to consider is whether the high level of C_{14} in prehistoric artifacts is a "smoking gun" left behind by a highly evolved civilization 10,000 years ago. As I have argued, a complete interpretation of a civilization such as ours is beyond the scope of one individual or group of individuals who are trained in only one discipline. Archaeologists and Egyptologists have interpreted and explained artifacts surviving ancient civilizations from a perspective that has resulted in a belief that our own civilization is the first to develop technology that uses electricity as a means of performing work. Working from this premise, it is not surprising that evidence such as the granite artifacts found in Egypt, which demand that we include the possibility of advanced technological knowledge existing in prehistory, has been misinterpreted, disregarded, or overlooked.

We also must consider, however, that if this unthinkable nuclear catastrophe actually transpired, someone would have put into writing the horror they witnessed. It is possible that such writings would survive the centuries to provide future historians with some clues to the horrific events, assuming those records were interpreted correctly. Without doubt, an event of such magnitude would leave its mark. And indeed, written records do entice us with clues of what could have been an ancient nuclear accident—or even an ancient nuclear war.

The ancient Indian Sanskrit text *The Mahābhārata* is a work that has no precise chronological origin. It is estimated that it was written around 400 B.C. but probably was copied from earlier texts from a much earlier date. A complete translation in eleven volumes, though unelegant in some scholars' minds, was made by Kesari Mohan Ganguli and published under the name P. Chandra Roy between 1883 and 1896.[5] The work is replete with references to terrible wars that involved the use of weapons that we normally do not associate with the primitive warriors of prehistory. The writer, or writers, of *The Mahābhārata* seemed to exaggerate, or get confused, when describing weapons that—given the era in which they were used—should have been limited to swords, spears, and bows and arrows. Was it imagination or wishful thinking that prompted the writer(s) to describe weapons that included missiles and "birds" that swooped down from the heavens, issuing forth fire to demolish entire forests? There also was a terrifying device that moved in a way that, if considered to be a simple projectile, defied the laws of physics:

> *Thus the terrifying tumult of war was rampant when the Gods Nara and Nārāyana joined the battle. The blessed Lord Visnu, upon seeing the divine bow in Nara's hand, called up with his mind his Dānava-destroying discus. No sooner thought-of than the enemy-burning discus appeared from the sky in a blaze of light matching the sun's, with its razor-sharp circular edge, the discus Sudarsana, terrible, invincible, supreme. And when the fiercely blazing, terror-spreading weapon had come to hand, God Acyuta [Visnu] with arms like elephant trunks loosed it, and it zigzagged fast as a flash in a blur of light, razing the enemy's strongholds. Effulgent like the Fire of Doomsday, it felled foe after foe, impetuously tearing asunder thousands of Dānavas and Daityas as the hand of the greatest of men let go of it in the battle. Here it was ablaze licking like a fire, there it cut down with a vehemence the forces of the Asuras. How it was hurled into the sky, then into the ground, and like a ghoul it drank blood in that war.[6]*

There seem to be forces at work in this battle that we do not possess even today. There is an intelligence that guides this discus. Is this intelligence just the imagination of the writer, or is it the report of an eyewitness observation? In order to justify the latter, we have to consider not only the intelligence that guided this discus, but the source of its energy. As though to answer our question, the text later refers to the "Elixir" that brought an added dimension to the ancient Indian wars so that they more closely parallel our own: "When that grand bird had rid them all of life, he strode across them to look for the Elixir. He saw fire everywhere; blazing fiercely, it filled all the skies with its flames, burning hot and razor-sharp rays, and evil under the stirring of the wind."[7] Then as if to make an association between the Elixir and its use: "He saw, in front of the Elixir, an iron wheel with a honed edge

250

and sharp blades, which ran incessantly, bright like fire and sun. . . . And behind the wheel he saw two big snakes, shimmering like blazing fires, tongues darting like lightning, mouths blazing, eyes burning, looks venomous, no less powerful than gruesome, in a perpetual rage and fierce, that stood guard over the Elixir, their eyes ever-baleful and never blinking. Whomever either snake's eyes were to fall upon would turn to ashes."[8] This passage brings to mind the important role gasoline has played in modern war, not only as a weapon, but as fuel for vehicles. Could the Elixir have been the gasoline that fueled these ancient conflagrations?

Perhaps the foregoing is just an ancient myth that has no basis in reality, although there are more references to other weapons of war that are closer to home and that have more meaning today than they did when the Sanskrit was first translated: "The King of the Gods, beholding the rage of Phalguna, unleashed his own blazing missile, which streaked across the entire sky. Thereupon the Wind God, who dwells in the sky, thunderously shaking all the oceans, generated towering clouds that sent forth shafts of water."[9]

With missiles streaking through the air against an opposing force, it may not be so surprising to find that the ancient Indians used these missiles in much the same way as the United States in the Gulf War with the Patriot missile: " . . . Filled with anger and vindictiveness, Parasurama brought forth a mighty weapon of Brahmā. On my part, I produced the same excellent weapon of Brahmā in order to counter the effect of his weapon. Those two weapons of Brahmā met each other in mid-air, without being able to reach either Rama or myself. Around them a flame blazed forth, and living things were greatly afflicted thereby."[10] As though to indicate the power of these mighty missiles, the ancient storyteller(s) wrote, "Thus sped by that mighty warrior, the shaft endowed with the energy of the Sun caused all the points of the compass to blaze with light."[11]

Knowing that the energy of the sun comes from the fusion of hydrogen atoms, the thought of hydrogen bombs brings terrible visions of vast destruction, mushroom clouds, and insidious radiation wafting across the land. These visions are included in other books that reference *The Mahābhārata* as testimony of nuclear war in prehistory. In *We Are Not the First*, Andrew Tomas wrote: " 'A blazing missile possessed of the radiance of smokeless fire

251

was discharged. A thick gloom suddenly encompassed the heavens. Clouds roared into the higher air, showering blood. The world, scorched by the heat of that weapon, seemed to be in fever,' thus describes the *Drona Parva* a page of the unknown past of mankind. One can almost visualize the mushroom cloud of an atomic bomb explosion and atomic radiation. Another passage compares the detonation with a flare-up of *ten thousand suns.*"[12]

Frederick Soddy, British chemist and Nobel prize winner for his work on the origin and nature of isotopes, discerned a vastly different meaning in these words than his contemporaries. Regarding the ancient Indian scriptures in 1909, before the atomic age, he wrote: "Can we not read into them some justification for the belief that some former forgotten race of men attained not only to the knowledge we have so recently won, but also the power that is not yet ours?"[13] Soddy's work with British phycisist Ernest Rutherford added to our understanding of the atom and led to the splitting of its nucleus by Sir John D. Cockroft and Ernest T. Walton in 1932. Soddy believed that civilizations in the past were familiar with the awesome power contained within the atom and had suffered the consequences of its misuse. In 1910 he wrote in his book, *Radium*:

> *Some of the beliefs and legends bequeathed to us by antiquity are so universal and firmly established that we have become accustomed to consider them as being almost as ancient as humanity itself. Nevertheless, we are tempted to inquire how far the fact that some of these beliefs and legends have so many features in common is due to chance, and whether the similarity between them may not point to the existence of an ancient, totally unknown and unsuspected civilization of which all other traces have disappeared.*[14]

Tomas pointed out that a skeleton was discovered in India that had up to fifty times more radioactivity than normal. He also puzzled over a meeting he had with Pundit Kaniah Yogi. He wrote:

> *According to Pundit Kaniah Yogi of Ambattur, Madras, whom I met in India in 1996, the original time measurement of the Brahmins was sexagesimal, and he quoted the* Brihath Sathaka *and other San-*

skrit sources. In ancient times the day was divided into 60 kala, *each equal to 24 minutes, subdivided into 60* vikala, *each equal to 24 seconds. Then followed a further sixty-fold subdivision of time into* para tatpara, vitatpara, ima *and finally* kashta—*or 1/300,000,000 of a second. The Hindus have never been in a hurry and one wonders what use the Brahmins made of these fractions of a microsecond. While in India the author was told that the learned Brahmins were obliged to preserve this tradition from hoary antiquity but they themselves did not understand it.*

Is this reckoning of time a folk memory from a highly technological civilization? Without sensitive instruments the kashta *would be absolutely meaningless. It is significant that the* kashta, *or 3×10^8 second, is very close to the life-spans of certain mesons and hyperons. This fact supports the bold hypothesis that the science of nuclear physics is not new.*

The Varahamira Table, *dated B.C. 550, indicates even the size of the atom. The mathematical figure is fairly comparable with the actual size of the hydrogen atom.*[15]

The indications that nuclear war was once a reality on this planet and was suffered by a civilization that was equally advanced as or more advanced than our own may be supported by some and rejected by others. However, we can no longer ignore the factual evidence that a prehistoric civilization capable of developing advanced machining techniques once existed on this planet. The theory I have presented in this book is based purely on fact, and I trust that readers will evaluate the deductions I have drawn from these facts with open-mindedness and objectivity.

That said, I would like now to revisit one of those deductions, the one that suggests the Egyptians understood the properties of gravity. It has been speculated on more than one occasion, and by more than one person, that this ancient civilization had the technology to neutralize the effects of gravity. If this were true, then the technological tools Egyptologists look for as evidence that the Egyptians were not primitive, such as the wheel or specific machinery, might have never existed—because the Egyptians would not have needed them! The simple fact is that the tools and machines we find so necessary in our gravity-bound civilization would not have been needed in a society that was able to control gravity.

If we were to develop the technology to overcome gravity, the energy expenditure of the peoples of the world would be sharply curtailed. Along with our diminished need for energy, we would no longer require many other ancillary products of an advanced society. Huge oil refineries, tire manufacturers, large manufacturing plants churning out massive engines and vehicle transmissions, and hundreds of thousands of miles of highways would conceivably become obsolete.

The point I am trying to make is that when we study the past seeking evidence of a highly advanced culture, we should not expect to find objects that we associate with our own culture. Different cultures develop along different paths. This process occurs even over relatively short periods of time, especially when one society is isolated from others. For example, when the Allies went into Germany after Hitler's defeat, they found that after only twelve years of isolation German technology was being developed along lines vastly different from our own. Pauwels and Bergier wrote:

> When the War in Europe ended on May 8th, 1945, missions of investigation were immediately sent out to visit Germany after her defeat. Their reports have been published; the catalogue alone has 300 pages. Germany had only been separated from the world since 1933. In twelve years the technical evolution of the Reich developed along strangely divergent lines. Although the Germans were behindhand as regards the atomic bomb, they had perfected giant rockets unmatched by any in America or Russia. They may not have had radar, but they had perfected a system of infra-red ray detectors which were quite as effective. Though they did not invent silicones, they had developed an entirely new organic chemistry, based on the eight-ring carbon chain.
>
> ... They had rejected the theory of relativity and tended to neglect the quantum theory. ... they believed in the existence of eternal ice and that the planets and the stars were blocks of ice floating in space. If it has been possible for such wide divergencies to develop in the space of twelve years in our modern world, in spite of the exchange of ideas and mass communications, what view must one take

of the civilizations of the past? To what extent are our archaeologists qualified to judge the state of the sciences, techniques, philosophy and knowledge that distinguished, say, the Maya or Khmer civilizations?[16]

The distance between our civilization and the one that built the pyramids obviously is far greater than that which separated us from Hitler's Germany. Still, we seem compelled to explain everything, even prehistoric cultures, in terms of our own knowledge and experience. We are seldom satisfied with an incomplete picture of the subject we are studying, and so taking fragments from here and there, we tend to fill in the gaps using deductive reasoning. Deductions, however, are contradictory, so we must scrutinize the facts from which they are drawn, not picking and choosing only those that help our case, but including all the evidence, no matter how discomforting to our beliefs.

My theory is that the Great Pyramid was the ancient Egyptians' power plant. However radical the idea may seem, it is, in my mind, supported by hard archaeological evidence. The artifacts reveal that the ancient Egyptians used advanced machining methods, which supports the deduction that their civilization, and perhaps others, was technologically advanced. Nevertheless, even with the powerful evidence I have presented throughout this book, and the growing support for such ideas, there is still a mountain of evidence—or lack of it—that prevents this theory's total acceptance. I acknowledge this truth, and I am open to revising my power plant theory if another theory presents itself to explain all the anomalies in the ancient artifacts and pyramids I have examined to build my own case.

The knowledge needed to evaluate certain of these ancient artifacts was not available until very recently. Even today there may be numerous articles that we will not understand until we further develop our own technology. We cannot fathom technology that is unknown to us, and we seldom consider things that seem impossible to us. Petrie, though knowledgeable in engineering and surveying, could not be expected to know anything about ultrasonic machining; hence his amazement at the machining abilities of the ancient Egyptians. Even if he had been aware of this technology, the intellectual climate of his time may have precluded his considering the possibility that these methods were known to the ancient Egyptians. Quite

...ply, the greatest barrier to our understanding may not necessarily be knowledge. It may be attitude.

One of the most inconceivable events with which modern humans are faced is nuclear disaster. Though the threat of all-out nuclear war between the United States and the former Soviet Union has been greatly reduced, it is still possible that our civilization could be wiped from the face of the Earth by a few miscalculations in foreign policy, a reckless terrorist act, or an error or malfunction in our own nuclear weapons or devices—the ones supposedly protecting us from a premature reaction to a nonexistent threat. Could it happen? Most of us believe that we, as a species, are simply too smart for these possibilities to overtake us.

Has it happened before? Were the ancient Egyptians smart enough to ensure that their own civilization would endure? The greatest lessons regarding our own mortality may begin with the pyramids of Egypt, the strong evidence of advanced machining practiced by the ancient Egyptians, the geological and biological records, and the world's ancient sacred records—these are all pieces of a giant puzzle that so many of us are trying to piece together. I have hope that we will regain this lost knowledge and learn from the lessons of the far distant past in time to save our own society from the fate that likely befell advanced civilizations that came before us. And I hope that along with granting us the wisdom to survive, this knowledge also may provide us the means through which we can evolve—spiritually, intellectually, and technologically—into more than we have ever chanced to dream.

Appendix A

THE MECHANICAL METHODS OF THE PYRAMID BUILDERS

by
Sir William Flinders Petrie

From *Pyramids and Temples of Gizeh*, 74–78.

Author's Note: These pages from Petrie are reproduced exactly from the original, including punctuation style. I have included this Appendix so that the reader will better understand the context within which Petrie presents his evidence.

he methods employed by the Egyptians in cutting the hard stones which they so frequently worked, have long remained undetermined. Various suggestions have been made, some very impracticable ; but no actual proofs of the tools employed, or the manner of using them, have been obtained. From the examples of work which I was able to collect at Gizeh, and from various fixed objects of which I took casts, the solution of the questions so often asked seems now to have been found.

The typical methods of working hard stones—such as granite, diorite, basalt, etc.—was by means of bronze tools ; these were set with cutting points, far harder than the quartz which was operated on. The material of these cutting points is yet undetermined; but only five substances are possible— beryl, topaz, chrysoberyl, corundum or sapphire and diamond. The character of the work would certainly seem to point to diamond as being the cutting jewel ; and only the consideration of its rarity in general, and its absence from Egypt, interfere with this conclusion, and render the tough uncrystallized corundum the more likely material.

Many nations, both savage and civilized, are in the habit of cutting hard materials by means of a soft substance (as copper, wood, horn, etc.), with a hard powder supplied to it ; the powder sticks in the basis employed, and this being scraped over the stone to be cut, so wears it away. Many persons have therefore readily assumed (as I myself did at first) that this method must necessarily have been used by the Egyptians ; and that it would suffice to produce all the examples now collected. Such, however, is far from being the case ; though no doubt in alabaster, and other soft stones, this method was employed.

That the Egyptians were acquainted with a cutting jewel far harder than quartz, and that they used this jewel as a sharp pointed graver, is put beyond doubt by the diorite bowls with inscriptions of the fourth dynasty, of which I found fragments at Gizeh ; as well as the scratches on polished granite of Ptolemaic age at San. The hieroglyphs are incised, with a very free-cutting point ; they are not scraped nor ground out, but are ploughed through the diorite, with rough edges to the line. As the lines are only 1/150 inch wide (the figures being about .2 long), it is evident that the cutting point must have been harder than quartz ; and tough enough not to splinter when so fine an edge was being employed, probably only 1/200 inch wide. Parallel lines are graved only 1/30 inch apart from centre to centre.

We therefore need have no hesitation in allowing that the graving out of lines in hard stones by jewel points, was a well known art. And when we find on the surfaces of the saw-cuts in diorite, grooves as deep as 1/100 inch, it appears far more likely that such were produced by the jewel points in the saw than by any fortuitous rubbing about of a loose powder. And when, further, it is seen that these deep grooves are almost always regular and uniform in depth, and equidistant, their production by the successive cuts of the jewel teeth of a saw appears to be beyond question. The best examples of equidistance are the specimens of basalt No. 4, (Pl. viii.), and of diorite No. 12 ; in these the fluctuations are no more than such as always occur in the use of a saw by hand-power, whether worked in wood or in soft stone.

On the granite core, broken from a drill hole (No. 7), other features appear, which can only be explained by the use of fixed jewel points. Firstly, the grooves which run around it form a regular spiral, with no more interruption or waviness than is necessarily produced by the variations in the component crystals ; this spiral is truly symmetrical with the axis of the core. In one part a groove can be traced, with scarcely an interruption, for a length of four turns. Secondly, the grooves are as deep in the quartz as in the adjacent feldspar, and even rather deeper. If these were in any way produced by loose powder, they would be shallower in the harder substance—quartz ; whereas a fixed jewel point would be compelled to plough to the same depth in all the components ; and further, inasmuch as the quartz stands out slightly beyond the feldspar (owing to the latter being worn by general rubbing), the groove was left even less in depth on the feldspar than on the quartz. Thus, even if specimens with similarly deep grooves would be produced by a loose powder, the special features of this core would still show that fixed cutting points were the means here employed.

That the blades of the saws were of bronze, we know from the green staining on the sides of saw cuts, and on grains of sand left in a saw cut.

The forms of tools were straight saws, circular saws, tubular drills, and lathes.

The straight saws varied from .03 to .2 inch thick, according to the work ; the largest were 8 feet or more in length, as the cuts run lengthways on the Great Pyramid coffer, which is 7 feet 6 in. long. The examples of saw cuts figured in Pl. viii. are as follow. No. 1, from the end of the Great Pyramid coffer of granite, showing where the saw cut was run too deep in the stuff twice over, and backed out again. No. 2, a piece of syenite, picked up at Memphis ; showing cuts on four faces of it, and the breadth of the saw by a cut across the top of it. This probably was a waste piece from cutting out a statue in the rough. No. 3, a piece of basalt, showing a cut run askew, and abandoned, with the sawing dust and sand left in it ; a fragment from the sawing of the great basalt pavement on the East of the Great Pyramid. No. 4, another piece from the same pavement, showing regular and well-defined lines. No. 5, a slice of basalt from the same place, sawn on both sides and nearly sawn in two. No. 6, a slice of diorite bearing equidistant and regular grooves of circular arcs, parallel to one another ; these grooves have been nearly polished out by cross grinding, but still are visible. The only feasible explanation of this piece is that it was produced by a circular saw. The main examples of sawing at Gizeh are the blocks of the great basalt pavement, and the coffers of the Great, Second, and Third Pyramids,—the latter, unhappily, now lost.

Next the Egyptians adapted their sawing principle into a circular, instead of a rectilinear form, curving the blade round into a tube, which drilled out a circular groove by its rotation ; thus, by breaking away the cores left in the middle of such grooves, they were able to hollow out large holes with minimum of labour. These tubular drills vary from 1/4 inch to 5 inches in diameter, and from 1/30 to 1/5 thick. The smallest hole yet found in granite is 2 inches diameter, all the lesser holes being in limestone or alabaster, which was probably worked merely with tube and sand. A peculiar feature of these cores is that they are always tapered, and the holes are always enlarged towards the top. In the soft stones cut merely with loose powder, such a result would naturally be produced simply by the dead weight on the drill head, which forced it into the stone, not being truly balanced, and so always pulling the drill over to one side ; as it rotates, this would grind off material from the core and the hole. But in the granite core, No. 7, such an explanation is insufficient, since the deep cutting grooves are scored out quite as strongly in the tapered end as elsewhere ; and if the taper was merely produced by rubbing of powder, they would have been polished away, and certainly could not be equally deep in quartz as in feldspar. Hence we are driven to the conclusion that auxiliary cutting points were inserted along the side, as well as around the edges of the tube drill ; as no granite or diorite cores are known under two inches diameter, there would be no impossibility in setting such stones, working either through a hole in the opposite side of the drill, or by setting a stone in a hole cut through the drill, and leaving it to project both inside and outside the tube. Then a preponderance of the top weight to any side would tilt the drill so as to wear down the groove wider and wider, and thus enable the drill and the dust to be the more easily withdrawn from the groove. The examples of tube drilling on Pl, viii. are as follow:— No. 7, core in granite, found at Gizeh. No. 8, section of cast of a pivot hole in a lintel of the granite temple at Gizeh ; here the core being of tough hornblende, could not be entirely broken out, and remains to a length of .8 inch. No. 9, alabaster mortar, broken in course of manufacture, showing the core in place ; found at Kom Ahmar (lat. 28°5'), by Prof. Sayce, who kindly gave it to me to illustrate this subject. No. 10, the smallest core yet known, in alabaster ; this l owe to Dr. Grant Bey, who found it with others at Memphis. No. 11, marble eye for inlaying, with two tube-drill holes, one within the other ; showing the thickness of the small drills. No. 12, part of the side of a drill-hole in diorite, from Gizeh, remarkable for the depth and regularity of the grooves in it. No. 13, piece of limestone from Gizeh, showing how closely the holes were placed together in removing material by drilling ; the angle of junction shows that the groove of one hole just overlapped the groove of another, probably without touching the core of the adjacent hole ; thus the minimum of labour was required. The examples of tube drilling on a large scale are the great granite coffers, which were hollowed out by cutting rows of tube drill-holes just meeting, and then breaking out the cores and intermediate pieces ; the traces of this work may be seen in the inside of the Great Pyramid coffer, where two drill-holes have been run too deeply into the sides ; and on a fragment of a granite coffer with a similar error of work on it, which I picked up at Gizeh. At El Bersheh (lat. 27°42') there is still a larger example, where a platform of limestone rock has been dressed down, by cutting it away with tube drills about 18 inches diameter ; the circular grooves occasionally intersecting, prove that it was done merely to remove the rock.

The principle of rotating the tool was, for smaller objects, abandoned in favour of rotating the work ; and the lathe appears to have been as familiar an instrument in the fourth dynasty, as it is in modern workshops. The diorite bowls and vases of the Old Kingdom are frequently met with, and show great technical skill. One piece found at Gizeh, No. 14, shows that the method employed was true turning, and not any process of grinding, since the bowl has been knocked off its centring, recentred imperfectly, and the old turning not quite turned out ; thus there are two surfaces belonging to different centrings, and meeting in a cusp. Such an appearance could not be produced by any grinding or rubbing process which pressed on the surface. Another detail is shown by fragment No. 15 ; here the curves of the bowl are spherical, and must have therefore been cut by a tool sweeping an arc from a fixed centre while the bowl rotated. This centre or hinging of the tool was in the apex of the lathe for the general surface of the bowl, right up to the edge of it ; but as a lip was wanted, the centring of the tool was shifted, but with exactly the same radius of its arc, and a fresh cut made to leave a lip to the bowl. That this was certainly not a chance result of hand-work is shown, not only by the exact circularity of the curves, and their equality, but also by the cusp left where they meet. This has not been at all rounded off, as would certainly be the case in hand-work, and it is clear proof of the rigid mechanical method of striking curves.

Hand graving tools were also used for working on the irregular surfaces of statuary ; as may well be seen on the diorite statue of Khafra found
at Gizeh, and now at Bulak.

The amount of pressure, shown by the rapidity with which the drills
and saws pierced through the hard stones, is very surprising ; probably a
load of at least a ton or two was placed on the 4-inch drills cutting in granite.
On the granite core, No. 7, the spiral of the cut sinks .1 inch in the circumference of 6 inches, or 1 in 60, a rate of ploughing out the quartz and feldspar which is astonishing. Yet these grooves cannot be due to the mere scratch-
ing produced in withdrawing the drill, as has been suggested, since there
would be about 1/10 inch thick of dust between the drill and the core at that
part ; thus there could scarcely be any pressure applied sideways, and the
point of contact of the drill and granite could not travel around the granite
however the drill might be turned about. Hence these rapid spiral grooves
cannot be ascribed to anything but the descent of the drill into the granite
under enormous pressure ; unless, indeed, we supposed a separate rymering
tool to have been employed alternately with the drill for enlarging the groove,
for which there is no adequate evidence.

That no remains of these saws or tubular drills have yet been found is
to be expected, since we have not yet found even waste specimens of work to
a tenth of the amount that a single tool would produce ; and the tool, instead of being thrown away like the waste, would be most carefully guarded.
Again, even of common masons' chisels, there are probably not a dozen
known ; and yet they would be far commoner than jeweled tools, and more
likely to be lost, or to be buried with the workman. The great saws and drills
of the Pyramid workers would be royal property, and would, perhaps, cost a
man his life if he lost one ; while the bronze would be remelted, and the
jewels reset, when the tools became worn, so that no worn-out tool would
be thrown away.

Appendix B

WROUGHT IRON FOUND IN THE GREAT PYRAMID

 as copper the only metal available to the ancient Egyptians? Notwithstanding the fact that cutting granite with copper chisels is an impossibility, Egyptologists have asserted that the pyramid builders predated the Bronze Age, and, therefore, were limited in their choice of metals with which to make their tools. Therefore, they say that copper was the only metal that the ancient Egyptians used to fashion the stones with which they built the Great Pyramid. They say this while evidence of prehistoric iron—proving that the ancient Egyptians had developed and used it when building the Great Pyramid—is in the keeping of the British Museum. The discoverers of this piece of iron go to great lengths to argue for and document its authenticity, as John and Morton Edgar point out in their book *Great Pyramid Passages*:

> *It is significant to note, in this connection, that a piece of wrought-iron was found in the Great Pyramid by one of Col. Howard Vyse's assistants, Mr. J.R. Hill, during the operations carried out at Giza in 1837. Mr. Hill found it embedded in the cement in an inner joint, while removing some of the masonry preparatory to clearing the southern air-channel of the King's Chamber. This piece of iron is probably the oldest specimen in existence; and Col. Howard Vyse was fully recognizant of the importance of the find. He forwarded it to the British Museum with the following certificates:*
>
> *'This is to certify that the piece of iron found by me near the (outside) mouth of the air passage, in the southern side of the Great Pyramid at Giza, on Friday, May 26th, was taken out by me from an inner joint, after having removed by blasting the outer two tiers of the stones of the present surface of the pyramid; and that no joint or opening of any sort was connected with the above-mentioned joint,*

by which the iron could have been placed in it after the original build-
ing of the Pyramid. I also showed the exact spot to Mr. Perring, on
Saturday, June 24th.—J. R. Hill'

'*To the above certificate of Mr. Hill, I can add, that since I saw*
the spot at the commencement of blasting, there have been two tiers
of stone removed, and that, if the piece of iron was found in the joint,
pointed out to me by Mr. Hill, and which was covered by a larger
stone partly remaining, it is impossible it could have been placed there
since the building of the pyramid.—J. S. Perring, C. E.'

'*We hereby certify, that we examined the place whence the iron*
in question was taken by Mr. Hill, and we are of the opinion, that the
iron must have been left in the joint during the building of the Pyra-
mid, and that it could not have been inserted afterwards.—Ed. S.
Andrews,—James Mash, C. E.' [1]

Despite the above testimonials, because the chronology for the development of metals did not include wrought iron in the age of the pyramids,
the specialists at the British Museum concluded that this wrought-iron artifact could not be genuine and must have been introduced in modern times.
Nevertheless, after examining the piece in 1881, Petrie objectively noted:

That sheet iron was employed we know, from the fragment found by
Howard Vyse in the masonry of the south air channel ; and though
some doubt has been thrown on the piece merely from its rarity, yet
the vouchers for it are very precise ; and it has a cast of nummulite on
the rust of it, proving it to have been buried for ages beside a block of
nummulitic limestone, and therefore to be certainly ancient. No rea-
sonable doubt can therefore exist about its being a really genuine piece
used by the pyramid masons ; and probably such pieces were required
to prevent crowbars biting into the stones, and to ease the action of
the rollers. [2]

Because of the British Museum's proclamation, and despite Petrie's
opinion, this metal plate received little attention until very recently when
Robert Bauval and Graham Hancock doggedly researched its history. They

reported that "Despite this forceful opinion from one of the oddball giants [Petrie] of Egyptology in the late Victorian Age, the profession as a whole has been unable to cope with the idea of a piece of wrought iron being contemporary with the Great Pyramid. Such a notion goes completely against the grain of every preconception that Egyptologists internalize throughout their careers concerning the ways in which civilizations evolve and develop."[3]

Hancock and Bauval go on to say that in 1989, after rigorous testing of a fragment cut from the plate, two eminent metallurgists, Dr. M. P. Jones, senior tutor in the Mineral Resources Engineering Department at Imperial College, London, and Dr. Sayed El Gayer, who gained his Ph.D. in extraction metallurgy at the University of Aston in Birmingham, reported that "it is concluded, on the basis of the present investigation, that the iron plate is very ancient. Furthermore, the metallurgical evidence supports the archaeological evidence which suggests that the plate was incorporated within the Pyramid at the time that structure was being built."[4]

Jones and El Gayer determined that the plate was not of meteoric origin and that it must have been smelted at between 1,000 and 1,100 degrees centigrade. They also discovered the presence of gold on one surface of the plate during these tests.

Armed with this expert data—and 110 years after Petrie's objective analysis—Hancock and Bauval spoke with Dr. A. J. Spencer and Dr. Paul Craddock of the British Museum, who characterized Jones' and El Gayer's conclusions as being "highly dubious," though they would not comment further to support this statement.

Having worked with metallurgists and technologists, and having read the works of and seen the documentaries of Egyptologists, when I compare these two conflicting opinions I place more trust in the science and objectivity of the metallurgists. Egyptologists have a vested interest in continuing their teachings as they have taught them for the past century. To do otherwise would be to admit that they have been wrong. The iron plate is just a small, though significant, item in a large collection of anomalies that have been ignored or misinterpreted by many academics because they contradict their orthodox beliefs.

ENDNOTES

INTRODUCTION

1   Peter Tompkins, *Secrets of the Great Pyramid*, 382.

2   William Fix, *Pyramid Odyssey*, 219.

CHAPTER ONE

1   Richard Hoagland at www.enterprisemission.com

2   Peter Tompkins, *Secrets of the Great Pyramid*, 218.

3   *Encyclopedia Britannica*, 15th ed., s.v. "Hegira."

4   Graham Hancock and Robert Bauval. *The Message of the Sphinx*, 101–103.

5   William Fix, *Pyramid Odyssey*, 65. The statement was made at the time Dr. Lehner was being funded by the Edgar Cayce foundation for his research in Egypt.

6   See Erich von Daniken, *Chariot of the Gods*.

7   Hancock, Bauval, and West are major proponents of the theory that the Sphinx was constructed during the age of Leo, over 10,000 years ago. West and geologist Robert Schoch claim that the weathering of the Sphinx and of the surrounding limestone wall is the result of repeated heavy rainfall, which has not occurred in Egypt for over 10,000 years.

8   Both Smyth quotations are from *The Great Pyramid: Its Secrets and Mysteries Revealed*, 132.

9   This is an estimate provided by Merle Booker, director of the Indiana Limestone Institute, to Richard Noone, author of *5/5/2000 Ice: The Ultimate Disaster*. Booker estimated that it would require that the entire Indiana limestone industry, comprised of thirty-three quarries, triple its current output, and the estimate did not allow for equipment breakdowns or labor problems.

10 Fix, *Pyramid Odyssey*, 66.

11 Ibid.

12 Kurt Mendelssohn, *The Riddle of the Pyramids*, 75.

13 I.E.S. Edwards, *Ancient Egypt: Discovering Its Splendors*, 84.

14 These two quotations are from *Ancient Egypt*, 88.

15 From an interview with the *Sunday Telegraph* (London), January 1, 1995.

16 See Petrie's *Pyramids and Temples of Gizeh*.

17 Peter Lemesurier, *The Great Pyramid Decoded*, 174–175.

18 Mark Lehner, *The Complete Pyramids*, 111.

19 Lehner's *The Complete Pyramid is* a marvelously executed graphic exploration of the pyramids of Egypt. To completely analyze each of the remarkably complex and mechanical designs found within each of these pyramids would require the same attention to each one as was given to the Great Pyramid for the past century and more.

20 Tompkins, 251. Architect James Hagan also agrees with these observations.

CHAPTER TWO

1 Peter Tompkins, *Secrets of the Great Pyramid*, 21.

2 The information regarding Greaves, Davison, Caviglia, and Davidson is recounted in Tompkins.

3 William Flinders Petrie, *Pyramids and Temples of Gizeh*, 19.

4 Tompkins, 249.

5 Ibid.

6 I.E.S. Edwards, *The Pyramids of Egypt*, 290–291.

7 Petrie, *Pyramids and Temples*, 86.

8 Ibid, 19.

9 Ibid, 26.

10 Edwards, *The Pyramids of Egypt*, 106.

11 Piazzi Smyth, *Our Inheritence in the Great Pyramid*, 174–175.

12 Ibid, 175–176.

13 Ibid, 174.

14 Petrie, *Pyramids and Temples*, 27.

15 Ibid.

16 Ibid.

17 Ibid, 28.

CHAPTER THREE

1 In manufacturing, "tolerance" is a unit of measure within which a deviation from a perfect form can be tolerated.

2 William F. Petrie, *Pyramids and Temples of Gizeh*, 13. Petrie's reference to the work of opticians has been taken out of context by modern writers. The accuracy required of some optical components is measurable to less than a wavelength of light.

3 Mark Lehner, *The Complete Pyramids*, 208–209.

4 Atlantis Rising Video, *Technologies of the Gods*, 1998. Hagan quotes are from this source and personal conversations.

5 Richard Noone, *5/5/2000 Ice: The Ultimate Disaster,* 105.

6 Petrie, *Pyramids and Temples,* 77.

7 Ibid, 13.

8 Max Toth, *Pyramid Prophecies,* 81.

9 I.E.S. Edwards, *Ancient Egypt,* 89.

10 Lehner, 209.

11 Netherlands Television, *Genesis in Stone,* 1995.

CHAPTER FOUR

1 I.E.S. Edwards, *Ancient Egypt,* 91.

2 William Flinders Petrie, *Pyramids and Temples of Gizeh,* 29.

3 Ibid, 35.

4 Ibid, 29.

5 Ibid.

6 Ibid, 77.

7 Ibid, 78.

8 Ibid, 75, 76, 78.

9 A. Lucas, *Ancient Egyptian Materials and Industries,* 88.

10 *Encyclopedia Britannica,* 15th ed., s.v. "Ultrasonics and Infrasonics."

CHAPTER FIVE

1 William Flinders Petrie, *Pyramids and Temples of Gizeh,* 35, 36.

2 I.E.S. Edwards, *The Pyramids of Egypt,* 134.

3 Petrie, *Pyramids and Temples,* xii.

4 Joseph Davidovits and Margie Morris, *The Pyramids: An Enigma Solved,* 85–86.

CHAPTER SIX

1 Information on Coral Castle can be obtained from Coral Castle, 28655 S. Dixie Highway, Homestead, FL 33030 USA. Tel: 305-248-6344. For a nominal charge, the proprietors will mail a package that includes Leedskalnin's papers on electricity and magnetism and a small self-published book entitled *A Book In Every Home,* in which Leedskalnin instructs the reader on how to raise a young lady. Is this a metaphor? I don't know; it is a strange little book and I have not quite figured it out yet.

CHAPTER EIGHT

1   *Encyclopedia Britannica*, 15th ed., s.v. "Sound, Sources of."

2   From the archive of "Dr. Magneto's Questions and Answers" at http://image.gsfc.nasa.gov/~odenwald/ask/q768.html. Other references on Earth's resonance can be found on the Internet by doing a search for "schumann resonance."

3   Quoted in Peter Tompkins, *Secrets of the Great Pyramid*, 72. Further Taylor quotes from the same source.

4   Ibid, 74.

5   Piazzi Smyth, *The Great Pyramid: Its Secrets and Mysteries Revealed*, 285.

6   Ibid, 287.

7   William Flinders Petrie, *Pyramids and Temples of Gizeh*, 13.

8   William Fix, *Pyramid Odyssey*, 232.

9   *Encyclopedia Britannica*, 15th ed., s.v. "Vibration."

10 Said may have meant "four *resonant* frequencies."

11 Quoted from Said's website at http://www.lauralee.com/said.htm (October 1997).

12 Paul Horn, *Inside the Great Pyramid*, Mushroom Records, 1977.

13 Tompkins, 279.

14 Bill Kerrel and Kathy Goggin, *The Guide to Pyramid Energy*, 14.

15 Bill Schul and Ed Pettit, *The Secret Power of Pyramids*, 105.

16 Dale Pond and Walter Baumgartner, *Nikola Tesla's Earthquake Machine*, 5–6.

17 Thomas Bearden, *Establishing a Standing Scalar EM Wave in the Earth*, 6–47.

18 Ibid, 6–45.

CHAPTER NINE

1   Dee Jay Nelson and David H. Coville, *Life Force in the Great Pyramids*, 73.

2   William Flinders Petrie, *Pyramids and Temples of Gizeh*, 28.

3   Ibid, 30.

4   Piazzi Smyth, *The Great Pyramid: Its Secrets and Mysteries Revealed*, 7.

5   Petrie, *Pyramids and Temples*, 31.

6   Paul Horn, *Inside the Great Pyramid*, Mushroom Records, 1977.

7   Ibid.

8   Petrie, *Pyramids and Temples*, 30.

9   Smyth, 447–448, 451, 452.

10 Ibid, 452.

11 Petrie, *Pyramids and Temples*, 8.

12 *Encyclopedia Britannica*, 15th ed., s.v. "Vibrations: Energy and Power in Vibrations."

13 Graham Hancock, *Fingerprints of the Gods*, 333.

14 Ibid.

15 Petrie, *Pyramids and Temples*, 21.

CHAPTER TEN

1   *Encyclopedia Britannica*, 15th ed., s.v. "Radio Sources, Astronomical."

2   Ibid, s.v. "Lasers and Masers."

3   Ibid, s.v. "Masers."

4   Piazzi Smyth, *The Great Pyramid: Its Secrets and Mysteries Revealed*, 416.

5   Ibid.

6   Ibid, 156.

CHAPTER ELEVEN

1   *Encyclopedia Britannica*, 15th ed., s.v. "Catalysis."

2   Graham Hancock and Robert Bauval, *Message of the Sphinx*, 115–116.

3   Richard Noone, *5/5/2000: Ice: The Ultimate Disaster*, 99.

4   Report prepared for me by chemical engineer Joseph Drejewski.

5   Hancock and Bauval, *Message of the Sphinx*, 51.

CHAPTER TWELVE

1   William Flinders Petrie, *Pyramids and Temples of Gizeh*, 87.

2   Peter Tompkins, *Secrets of the Great Pyramid*, 249.

CHAPTER THIRTEEN

1   Peter Tompkins, *Secrets of the Great Pyramid*, 107.

CHAPTER FOURTEEN

1   Nikola Tesla, *The Fantastic Inventions of Nikola Tesla*, 219–240.

2   Brad Steiger, *Worlds Before Our Own*, 74–75.

3 Ibid, 75–76.

4 Robert Bauval and Adrian Gilbert, *The Orion Mystery*, 73–74.

5 Ibid, 90–91.

6 Edgar Evans Cayce, *Edgar Cayce on Atlantis*, 88–90.

7 *The New Webster Encyclopedic Dictionary of the English Language*, 1973, s.v., "igneous."

8 Cayce, *Edgar Cayce on Atlantis*, 85.

CHAPTER FIFTEEN

1 Brad Steiger, *Worlds Before Our Own*, 16.

2 David Wilson, *The New Archaeology*, 97.

3 Willard Frank Libby Ph.D. received the 1960 Nobel Prize for chemistry for his work on radioactive carbon dating.

4 Wilson, 112.

5 J.A.B. van Buitenen, *The Mahābhārata*, xxv–xxxvii.

6 Ibid, 75–76.

7 Ibid, 88.

8 Ibid, 89.

9 Ibid, 419.

10 Chakravarthi V. Narasimhan, *The Mahābhārata*, 117–118.

11 Ibid, 166.

12 Andrew Tomas, *We Are Not the First*, 75.

13 Ibid.

14 Quoted in Louis Pauwels and Jacques Bergier, *The Morning of the Magicians*, 181.

15 Tomas, *We Are Not the First*, 76.

16 Pauwels and Bergier, *The Morning of the Magicians*, 170–171.

APPENDIX B

1 Quoted in Max Toth, *Pyramid Prophecies*, 208–209.

2 William Flinders Petrie, *Pyramids and Temples of Gizeh*, 85.

3 Graham Hancock and Robert Bauval, *The Message of the Sphinx*, 106.

4 Ibid, 106–107.

BIBLIOGRAPHY

Bauval, Robert, and Adrian Gilbert. *The Orion Mystery: Unlocking the Secrets of the Pyramids*. New York: Crown, 1994.

Bearden, Thomas. "Maxwell's Lost Unified Field Theory." In *Proceedings of the International Tesla Symposium*. Edited by Steven R. Elswick, section 6, 24–68. Colorado Springs, CO: International Tesla Society, 1988.

Cayce, Edgar Evans. *Edgar Cayce on Atlantis*. New York: Warner Books, 1968.

Davidovits, Joseph, and Margie Morris. *The Pyramids: An Enigma Solved*. New York: Hippocrene Books, 1988.

Edwards, I.E.S. "Pyramids: Building for Eternity." In *Ancient Egypt: Discovering Its Splendors*. Edited by Gilbert M. Grosvenor, 73–101. Washington, D.C.: National Geographic Society, 1978.

———. *The Pyramids of Egypt*. London: Penguin Books, 1993.

Evans, Humphrey. *The Mystery of the Pyramids*. New York: Thomas Y. Crowell, 1979.

Fix, William. *Pyramid Odyssey*. New York: Mayflower Books, 1978.

Hancock, Graham. *Fingerprints of the Gods*. New York: Crown, 1995.

Hancock, Graham, and Robert Bauval. *The Message of the Sphinx*. New York: Crown, 1996.

Hodges, Henry. *Technology in the Ancient World*. New York: Barnes and Noble, 1992.

Horn, Paul. "Paul Horn—Inside the Great Pyramid." Booklet included in LP record of the same name. Los Angeles: Mushroom Records, 1977.

Kerrel, Bill, and Kathy Goggin. *The Guide to Pyramid Energy*. Santa Monica, CA: Pyramid Power - V, 1975.

Knudsen, Vern O. "Architectural Acoustics." In *Scientific American: The Physics of Music*. San Francisco: W.H. Freeman and Company, 1963.

Leedskalnin, Edward. *A Book in Every Home*. Homestead, FL: Edward Leedskalnin, 1936.

———. *Magnetic Current*. Homestead, FL: Edward Leedskalnin, 1945.

Lehner, Mark. *The Complete Pyramids: Solving the Ancient Mysteries*. London: Thames & Hudson, 1997.

Lemesurier, Peter. *The Great Pyramid Decoded*. New York: Avon Books, 1979.

Lucas, A. *Ancient Egyptian Materials and Industries*. London: Histories and Mysteries of Man, 1989.

The Mahābhārata: 1. The Book of the Beginning. Translated and edited by J.A.B van Buitenen. Chicago: University of Chicago Press, 1983.

The Mahābhārata. Translated by Chakravarthi V. Narasimhan. New York: Columbia University Press, 1965.

Mendelssohn, Kurt. *The Riddle of the Pyramids.* New York: Praeger, 1974.

Nelson, Dee Jay, and David Coville. *Life Force in the Great Pyramid.* Marina del Rey, CA: DeVorss & Company, 1977.

Noone, Richard. *5/5/2000 Ice: The Ultimate Disaster.* New York: Harmony Books, 1986.

Pauwels, Louis, and Jacques Bergier. *The Morning of the Magicians.* New York: Avon, 1968.

Petrie, William Flinders. *Pyramids and Temples of Gizeh.* 1883. Reprint, with an update by Dr. Zahi Hawass, London: Histories and Mysteries of Man, 1990.

____. *Ten Years Digging in Egypt.* 1891. Reprint, Chicago: Ares Publishers, 1989.

Pond, Dale, and Walter Baumgartner. *Nikola Tesla's Earthquake Machine.* Santa Fe, NM: The Message Company, 1995.

Schul, Bill, and Ed Pettit. *The Secret Power of Pyramids.* Greenwich, CT: Fawcett Publications, 1975.

Smyth, Piazzi. *The Great Pyramid: Its Secrets and Mysteries Revealed.* 4th. ed. New York: Bell Publishing Company, 1978. Originally published as *Our Inheritance in the Great Pyramid.* London: W. Isbister, 1880.

Sprague de Camp, Lyon. *The Ancient Engineers.* New York: Ballentine Books, 1974.

Steiger, Brad. *Worlds Before Our Own.* New York: Berkley Publishing, 1978.

Technologies of the Gods. Produced and directed by J. Douglas Kenyon and N. Thomas Miller. 65 min. Atlantis Rising Video, 1998. Videocassette.

Tesla, Nikola, and David Hatcher Childress. *The Fantastic Inventions of Nikola Tesla.* Kempton, IL: Adventures Unlimited, 1993.

Tomas, Andrew. *We Are Not the First: Riddles of Ancient Science.* London: Souvenir Press, 1971.

Tompkins, Peter. *Secrets of the Great Pyramid.* New York: Harper & Row, 1971.

Wilson, Colin. *From Atlantis to the Sphinx.* London: Virgin Books, 1996.

Wilson, David. *The New Archaeology.* New York: Alfred A. Knopf, 1975.

The World's Last Mysteries. New York: Reader's Digest Association, 1978.

INDEX

ABOUT THE AUTHOR

 hristopher Dunn has an extensive background as a master crafts-man, starting as a journeyman lathe turner in his home town of Manchester, England. Recruited by an American aerospace company, he immigrated to the United States in 1969. Beginning as a skilled machinist, he has worked at every level of high-tech manufacturing from toolmaking to operating high-power industrial lasers, including the position of Project Engineer and Laser Operations Manager at DMS, a Midwest aerospace manufacturer.

The author's pyramid odyssey began in 1977 when he read Peter Tompkins' book *Secrets of the Great Pyramid*. His immediate reaction to the Giza Pyramid's schematics was that this edifice was a gigantic machine. Discovering the purpose of this machine and documenting his case has taken the better part of twenty years of research. In the process he has published a dozen magazine articles, including the much quoted *Advanced Machining in Ancient Egypt* in *Analog*, and has had his research referenced in such books as Graham Hancock's *Fingerprints of the Gods* and Colin Wilson's *From Atlantis to the Sphinx*. Chris Dunn, his wife Jeanne, and their children live in Danville, Illinois.

Christopher Dunn is featured in the Mystic Fire video *Technologies of the Gods: The Case for Pre-Historic High Technology*, which brings together convincing evidence that ancient civilizations utilized high-tech engineering methods equal to if not superior to our own and that these technologies were being applied on a worldwide level. This case is presented by such renowned experts as Robert Bauval, Richard Noone, Colin Wilson, John Michell, Patrick Flanagan, Zecharia Sitchin, David Hatcher Childress, Edgar Evans Cayce, as well as Christopher Dunn. You can order *Technologies of the Gods* by writing Mystic Fire Video, P.O. Box 422, Prince Street Station, New York, NY 10012, or phone 1-800-292-9001.

BOOKS OF RELATED INTEREST
BY BEAR & COMPANY

CATACLYSM!
Compelling Evidence of a Cosmic Catastrophe in 9500 B.C.
by D.S. Allan and J.B. Delair

MAYA COSMOGENESIS 2012
The True Meaning of the Maya Calendar End-date
by John Major Jenkins

THE HIDDEN MAYA
A New Understanding of Maya Glyphs
by Martin Brennan

THE MAYAN FACTOR
Path Beyond Technology
by José Argüelles

THE EGYPTIAN ORACLE
by Maya Heath

THE MYSTERY OF THE CRYSTAL SKULLS
A Real-Life Detective Story of the Ancient World
by Chris Morton and Ceri Louise Thomas

SECRETS OF MAYAN SCIENCE/RELIGION
by Hunbatz Men

STARWALKING
Shamanic Practices for Traveling Into the Night Sky
by Page Bryant

Contact your local bookseller

—or—

Bear & Company
P.O. Box 2860
Santa Fe, NM 87504

1-800-WE-BEARS